GEOGRAPHICAL READINGS

Transport and Development

978 0333 144787

D1796972

Transport and Development

EDITED BY

B. S. HOYLE

MACMILLAN

Selection, editorial matter and introduction
© B. S. Hoyle 1973

All rights reserved. No part of this publication may be
reproduced or transmitted, in any form or by any means,
without permission

First published 1973 by
THE MACMILLAN PRESS LTD
London and Basingstoke
Associated companies in New York
Melbourne Dublin Johannesburg and Madras

SBN 333 14477 5 (hard cover)
SBN 333 14478 3 (paper cover)

Text set in 10/12 pt. Monotype Times New Roman, printed by letterpress,
and bound in Great Britain at The Pitman Press, Bath

Contents

Acknowledgements

'Transport Expansion in Underdeveloped Countries', by Edward J. Taaffe, Richard L. Morrill and Peter R. Gould, reprinted from the *Geographical Review*, **53** (1963) and 'Transport Expansion in Liberia', by William R. Stanley, reprinted from the *Geographical Review*, **60** (1970), copyrighted by the American Geographical Society of New York; 'Towards a Theory of Transport and Development', by George W. Wilson, from *The Impact of Highway Investment on Development*, by George W. Wilson, Barbara R. Bergmann, Leon V. Hirsch and Martin S. Klein © 1966, by the Brookings Institution, Washington D.C.; 'Recent Railway Construction in Tropical Africa', by A. M. O'Connor, from *Transport in Africa*, proceedings of a symposium held at the Centre of African Studies 1969, by permission of Centre of African Studies Committee, University of Edinburgh; 'Transportation and Urban Development in West Africa: A Review', by Shalom Reichman, from CNRS publications of papers from the International Conference on *La croissance urbaine en Afrique noire et à Madagascar*, by permission of Centre National de la Recherche Scientifique; 'Container Potential of West African Ports', by David Hilling, from *Dock and Harbour Authority*, **50,** No. 583 (1969), reprinted by permission of The Dock and Harbour Authority; 'Geography, Transportation and Regional Development', by Howard L. Gauthier, from *Economic Geography*, **46** (1970), reprinted by permission of Economic Geography; 'Highway Improvements and Agricultural Production: An Argentine Case Study', by F. Miller, from *Traffic Quarterly*, **22** (1968), reprinted by permission of Eno Foundation for Highway Traffic Control, Inc.; 'The Search for Spatial Regularities in the Development of Australian Seaports 1861–1961/2', by Peter J. Rimmer, from *Geografiska Annaler*, **49B** (1967), reprinted by permission of Geografiska Annaler; 'The Importance of Passenger Transport in Nigeria', by Alan Hay, *Nigerian Journal of Economic and Social Studies*, **11** (1969), by permission of Nigerian Journal of Economic and Social Studies (Nigerian Economic Society); 'Transportation and the Growth of the São Paulo

Economy', by H. L. Gauthier, from *Journal of Regional Science*, **8,** No. 1, reprinted by permission of the Journal of Regional Science; 'Transport and Economic Growth in Developing Countries: The Case of East Africa', by B. S. Hoyle, from *Geographical Essays in Honour of K. C. Edwards*, eds. R. H. Osborne, F. A. Barnes and J. C. Doornkamp (1970), reprinted by permission of the University of Nottingham.

Introduction

THE relationship between transport and development is a subject of considerable theoretical interest and practical importance, and one that has occupied a good deal of attention over many years in both advanced and less-developed countries. The interaction between the level and pattern of transport resources and the average level of living of the population of an area is a critical factor affecting economic and social progress, and must be taken into account at all stages of national and regional development planning. In the advanced countries, much attention was paid to transport innovation during the formative years of industrial growth; today, new strategies of economic planning require the modification or renewal of inherited transport systems. In the less-developed countries, to which this volume primarily refers, there is widespread concern for transport in the context of the desire to promote rapid economic development. The spectrum of transport modes available in the less-developed world ranges from head porterage to jumbo jets and from canoes to containers, although there may be no intermodal choice in one locality. Yet 'there is no escape from transport' (Munby, 1968); even in the most remote and least developed of inhabited regions, transport in some form is a fundamental part of the daily rhythm of life.

The study of transport is not the unique prerogative of any one discipline, but is shared by many fields of inquiry each with its own range of viewpoints. Economists are concerned particularly with the assessment of the demand for transport and with the problems of the cost of overcoming distance (Fromm, 1965; Prest, 1969); the costs of investment in the transport sector are weighed objectively against the benefits likely to accrue, and against the potential feedback from investment in alternative sectors. For the geographer the chief importance of transport arises from its role as one of the principal factors affecting the location and distribution of economic and social activities. Geographers are thus concerned with the changing spatial structure of transport networks in relation to other aspects of the landscape, and with the factors affecting changing patterns of traffic flow. Studies have emphasised the importance of the morphological approach to the geography of transport (Appleton, 1965) in the context of land use and economic history; and in recent years the increasing use of models and quantitative techniques has brought a greater degree of precision to the analysis of transport networks. Together with colleagues in other disciplines, both economists and geographers are concerned with the development impact of transport improvements, and have attempted to assess relationships between changes in the transport sector and the evolving pattern of economic development within the area served. It is these relationships which the material brought together in this volume seeks to explore and to illustrate.

Ideas about the nature of the relationship between transport and development have changed considerably over time. Of course, the basic importance of transport is not in question: transport is clearly a factor of fundamental importance in all economic activity, and the cost of transport one of the most significant variables in the market price of any commodity. But beyond the basic level of infrastructural provision, where transport (like labour, capital, markets, land and power supplies) is an obvious *sine qua non* for modern economic growth, it quickly becomes a matter for debate and inquiry whether, as development proceeds, it is advantageous to extend or otherwise improve transport facilities, or whether limited capital resources available for investment might more efficiently and beneficially be used in other ways. This is a matter of vital concern to development planners, and it is important to maintain an awareness of the multi-dimensional nature of the problem: the economic, social, political and spatial dimensions of transport are all important and in some respects complementary—although it may be argued that it is frequently the political dimension in which particular situations and problems are predominantly viewed. The transport sector may, with some justification, be regarded as an epitome of relationships between terrain, economic history, social and political systems, and levels and patterns of development.

Much of the literature of transport economics and of the transport geography of less-developed countries reveals a widespread belief in the importance of transport and in the efficiency of transport improvements in accelerating the development process. Perhaps the most extreme claim for transport was that made by the colonial administrator Lord Lugard, who wrote over fifty years ago that 'the material development of Africa may be summed up in the one word —transport' (Lugard, 1922). A more recent United Nations study claimed that transport is 'the formative power of economic growth, and the differentiating process (Voigt, 1967); but it is also sometimes stated that transport is a result, rather than a cause, of economic development. Such statements are of necessity gross oversimplifications. The transport/development relationship is essentially a two-way interaction process, and the results of the interaction depend upon the type of economy involved and upon the level of development at which transport improvements are effected. At a given stage of development, an area requires a certain level of transport provision in order to maximise its potential; there is thus an *optimum transport capacity* for any development level. The existence of unsatisfied demand for transport may, over time, have serious adverse effects on the economy; conversely, the results of over-capitalisation may be unpleasant if too much is spent on transport provision in anticipation of demand which never materialises.

Both economists and geographers have made important contributions to the theory and methodology of transport studies, and in recent years an increasing degree of interdependence between the two disciplines has begun to emerge. In the 1950s relatively little attention was paid to transport by either discipline, and each appeared to approach the subject from the aspect furthest removed from the other. Thus transport geography was concerned with the effects of economic activities on transport patterns, with explanations of traffic flows, and with descriptions of transport systems; transport economics, on the other hand, was concerned almost entirely with 'matters of organisation, competition and charging, rather than with the effects of transport facilities on economic activities' (O'Connor, 1965*a*). More recently there has been an increasing recognition by economists of the importance of the spatial variable, and by geographers of the value of economic models and generalisations. Both disciplines have moved away from the unquestioning assumption that transport automatically promotes development and have shown that transport constitutes one element in a varied infrastructure necessary for economic growth, no more and no less important than other elements, and that it does not provide necessarily the positive stimulus which many had previously assumed. Emphasis on the permissive element in transport provision has led to a further viewpoint which maintains that transport investment may have a negative impact on economic growth, in that the creation or expansion of transport capacity may absorb scarce resources which could be more productively employed in other ways.

The objective evaluation of investment alternatives thus emerges as a major contribution of economists to the study of transport and development. It has been suggested, however, that in much of the less-developed world the economic models available are either too general for local development planners, below the level of large inter-urban connections, or too specific, dependent upon data in a form and at a level of precision generally unavailable in low-income countries (Birdsall, 1971). There is some truth in this criticism, but a wide range of material has been provided, for example, by the Brookings Institution (Washington D.C.) through its Transport Research Program which during the 1960s yielded a number of important studies of the role of transport in development; these include a perceptive general volume by the Program's Director, Wilfred Owen (Owen, 1964), studies of highway development in a range of less-developed countries (Wilson *et al.*, 1966), an analysis of relationships between transport and economic integration in South America (Brown, 1966), and a study of transport problems in India (Owen, 1968). In these and other studies originating in the Brookings Program, numerous writers have sought to explore and to illustrate economic aspects of the transport/development relationship, and the authors whose work is reproduced in the present volume owe a considerable debt to the Brookings research. Geographers, too, have long recognised that economic development is closely related to the growth of efficient transport systems, and have evolved numerous methods of spatial analysis in this context in recent years. The use of graph theory as a method of transport network analysis has received considerable attention; introduced by Garrison (1960) and elaborated by Kansky (1963), network analysis now has a wider significance in geography as a whole (Haggett and Chorley, 1969). Graph theoretical analysis provides 'perhaps the only precise means of analysing the total accessibility and linkage within the network, and . . . represents a significant step forward in the analysis of transport patterns' (Ward, 1969). An attempt to relate transport network densities to development levels in statistical terms was made by Berry (1960) who used several measures of transportation among his criteria for the construction of an economic-demographic development scale; discussions of this technique (for example Haggett, 1965) explore the relationships between high and low road and rail densities and the relatively high- or low-ranking position of countries on the scale, and suggest reasons for some anomalous cases. Another well-known attempt to analyse relationships between transport provision and the process of economic development is the study by Taaffe *et al.* (1963) based on West African material (Gould, 1960), and reproduced in part in this volume. The ideal-typical sequence of transport network development in a less-developed country, postulated by Taaffe *et al.*, has provided a basis for much later work, and the attempt to identify factors helping to explain spatial variations in transport network densities is important as a pointer towards a model simulating the economic growth process and allowing the evaluation of alternative transport development strategies. Studies based in part on the model devised by Taaffe and his colleagues include the two papers by Hoyle (on East Africa) and by Rimmer (on Australia) included in this volume, the latter referring primarily to the formative period of Australian economic development. Reference is made in these studies to the importance of distinguishing between the development impact of transport provision in the earliest stages of economic development, and the anticipated effects when later elaborations are undertaken. In the initial stages of economic growth, the introduction of modern transport makes a wide variety of new economic opportunities available simultaneously, and is therefore likely to promote growth; in the more advanced stages, transport is one of many sectors in which productive investment may be channelled, and the observed effects of initial transport provision should not be falsely projected as likely results of modern transport elaboration.

Whilst much of the literature on transport and development is concerned with the transport sector as a whole, much attention has also been paid to specific modes and to intermodal competition (for example Hazlewood, 1964). Railways have often assumed a particular predominance in the transport systems of less-developed countries, because in many cases a railway has acted as a major initiator of development and modernisation and has tended to dominate subsequent patterns of growth both in the transport sector and elsewhere. An arterial railway is one of the more obvious characteristic features of many former colonial areas, and its economic role has to some extent been circumscribed by the political motivations which surrounded its initial establishment. Only in relatively recent years, as some countries have developed complementary modern road systems, and underused railways have become too expensive to maintain, has the degree of road–rail competition become so intensive that railways have ceased in some cases to be economically viable. In at least two tropical African cases—Sierra Leone and Ivory Coast—the traditional close association between an arterial railway and the economic geography of the country concerned has begun seriously to disintegrate, and in the former case the railway system is being phased out as a new road programme is developed. At the same time, elsewhere in Africa and in other parts of the less-developed world, new railways are being built; the redundant parts of dynamic systems are being excised, whilst new elaborations are introduced in the context of changing economic circumstances (Barbour, 1967; O'Connor, 1965*b*). This, particularly, involves the development of inter-regional links (such as the Tazara line in Eastern Africa), designed to supplement the more conventional links between core areas of export production and coastal ports.

A common theme in transport studies in less-developed countries is 'the steady rise in the importance of road traffic, which first complements the railroad, then competes with it, and finally overwhelms it' (Taaffe *et al.*, *Geographical Review*, 53, p. 514). This process seems to occur at various stages of development, and major roads are probably being built today where rail arteries would have been constructed in the past. Although railways have fulfilled an important pioneer role in the establishment of a transport infrastructure in less-developed countries, more attention is now frequently paid to roads. In many such countries there already exists a country-wide if rather skeletal framework of roads, and the present process of road building and road improvement is one of strengthening and intensifying an existing system. It is generally recognised that the capacity and connectivity of a road network is a very important factor in regional and national development; one of the foremost priorities of transport policy in many less-developed countries is therefore the improvement of rural road systems, involving the application of development capital at the grass roots level of the transport sector in an attempt to bring at least one form of modern transport within reach of a high proportion of the population. Whilst transport development at this level may assist agricultural activity (Stanley, 1971) and the promotion of rural industries at an intermediate technological level, improved transport links between rural areas and urban-industrial growth centres may tend to increase the polarisation and exacerbate the problems of the towns. The diffusion of the opportunity to participate in an integrated national economy may be of great benefit, but improvements in road communications may produce quite different effects at different scales of operation (McMaster, 1970). The practical problems of road construction in the less-developed countries are numerous, and include the detailed analysis of the economic climate within which road developments are proposed, and the technical evaluation of the terrain involved. In this context, the Tropical Section of the Transport and Road Research Laboratory at Crowthorne, Berkshire, has done much valuable work including the development of important terrain evaluation techniques (Tingle, 1971).

The critical role of the seaport in the development process has received increased attention in recent years (Hoyle and Hilling, 1970), emphasising that a seaport in a less-developed country, through which almost all the external trade of that country passes, is well placed to act either as a growth pole or as a restrictive influence upon development. Which of these alternatives becomes the dominant trend depends partly upon the range, level and efficiency of the port services available and partly upon the transport policies of the relevant authorities (Peterec, 1967). A lack of adequate port facilities can be a major deterrent to national and regional economic growth, whilst the provision of modern port services may positively assist development by removing the inhibiting factor of low-level port facilities. For a variety of historical, geographical and economic reasons the less-developed countries are dependent upon overseas trade, and the expansion of trade is an essential prerequisite for modern economic growth. In this situation there is often a close relationship between the stage of economic development reached in a given less-developed country and the level and pattern of port services available in the area. The efficiency of port services in the less-developed countries is frequently, however, seriously affected by three important factors: the inheritance from the colonial period in terms of transport infrastructure and economic systems; the predominant role of primary tropical products in the trade structure of the less-developed countries; and the need to establish modern port systems at a time of rapidly changing world trading patterns and far-reaching changes in cargo-handling techniques. The basic dilemma of ports in less-developed countries is thus a question of balancing national systems, opportunities and goals with international trends in commodity flow and transport technology.

A wide variety of problems is thus raised by the question of transport and its role in development. The cost of overcoming the restrictive impact of distance is one of the basic factors affecting progress in the less-developed countries, and the promotion of increased mobility is everywhere a primary objective of development planning in the socio-economic sphere. The less-developed countries have progressed beyond a stage when almost any transport improvement seemed bound to bring economic benefit, to one in which the main need is not just for more and better transport facilities but rather for a greater degree of precision in the establishment of transport policies in the sight of some appreciation of what transport can and cannot be expected to do and of what else must be done if transport investment returns are to be maximised. Transport does not work miracles, and integration of transport planning with other economic sectors is essential to development strategy. Integration within the transport sector too is important; but at the grass roots level of transport, we should not lose sight of the fact that the most primitive modes—head porterage and canoes—account for the greatest proportion of goods movement in the less-developed countries. In view of the tremendous importance of transport of this sort in the lives of millions of ordinary people in tropical areas, a high priority of transport policy should surely be to extend and improve rural road systems so as to bring some form of modern transport within reach of most people and thus increase overall mobility. In a social and political context, in terms of improving the quality of life for the average man and woman in the bush, it may be claimed that this is the most important transport priority of all. But the transport problems of the less-developed countries should not be viewed in isolation from one another or from the rest of the world system of which they form a part. Transport media provide at best a framework for integration, and the power of transport to facilitate development and the promotion of inter-regional and international understanding should not be underestimated.

The writings reproduced in this volume are drawn from a fairly extensive body of literature available on the role of transport in the development of the less-developed countries. Related literature on other aspects of transport and development is more copious, and this volume is designed in part to encourage further reading within a wider framework. The selections presented here introduce a variety of problems and areas: material based on economic theory and on spatial analysis has been included; discussions of the role of specific transport modes—especially roads, railways and seaports—have been chosen; problems of intermodal integration, technological change and passenger transport are represented; and related issues involving agriculture, industry and urban growth are also discussed. Contributions have been selected to represent both general and particular problems within the field, and to draw upon work undertaken in many different parts of the less-developed world. Several of the articles included have been extensively revised for the present volume, others have been shortened, but most are reproduced substantially as originally printed. The Editor is grateful to the authors who have made their work available for this collection, and in particular wishes to thank Alan Hay, David Hilling and Tony O'Connor whose advice contributed towards the broad pattern of selection. Whilst the items here presented are mainly the work of geographers, the Editor hopes that the viewpoints expressed and the methods used in the material selected will be of interest to a wide range of those concerned with the dilemmas of development.

REFERENCES

APPLETON, J. H. (1965). *A Morphological Approach to the Geography of Transport*, University of Hull, Occasional Papers in Geography, 3.

BARBOUR, K. M. (1967). A survey of the Bornu railway extension in Nigeria—a geographical audit, *Journal of the Geographical Association of Nigeria*, **10**, 11–28.

BERRY, B. J. L. (1960). *An Inductive Approach to the Regionalization of Economic Development*, University of Chicago Department of Geography, Chicago, Research Paper, 62.

BIRDSALL, S. S. (1971). A model for determining road investments priorities in agriculturally underdeveloped areas, *East Lakes Geographer*, **7**, 60–70.

BROWN, R. T. (1966). *Transport and the Economic Integration of South America*, Washington.

FROMM, G. (ed.) (1965). *Transport Investment and Economic Development*, Washington.

GARRISON, W. L. (1960). Connectivity of the interstate highway system, Regional Science Association, *Papers and Proceedings*, **6**, 121–37.

GOULD, P. R. (1960). *Transportation in Ghana*, Evanston, Northwestern University Press, Studies in Geography, 5.

HAGGETT, P. (1965). *Locational Analysis in Human Geography*, London, pp. 76–79.

HAGGETT, P. and CHORLEY, R. J. (1969). *Networks in Geography*, London.

HAZLEWOOD, A. (1964). *Rail and Road in East Africa*, Oxford.

HOYLE, B. S. and HILLING, D. (eds.) (1970). *Seaports and Development in Tropical Africa*, London.

KANSKY, K. J. (1963). *Structure of Transportation Networks: Relationships between Network Geometry and Regional Characteristics*, University of Chicago Department of Geography Research Paper, Chicago, 84.

LUGARD, SIR F. D. (1922). *The Dual Mandate in British Tropical Africa*, Edinburgh, p. 5.

MCMASTER, D. N. (1970). Road communications and the pattern of rural settlement, in *Transport in Africa*, University of Edinburgh, Centre of African Studies, pp. 1–21.

MUNBY, D. (1968). *Transport*, Penguin, Harmondsworth, p. 7.

O'CONNOR, A. M. (1965a). *Railways and Development in Uganda*, Nairobi, Oxford University Press for the East African Institute of Social Research, p. 2.

O'CONNOR, A. M. (1965b). New railway construction and the pattern of economic development in East Africa, *Transactions of the Institute of British Geographers*, **36**, 21–30.

OWEN, W. (1964). *Strategy for Mobility*, Washington.

OWEN, W. (1968). *Distance and Development: Transport and Communications in India*, Washington.

PETEREC, R. J. (1967). *Dakar and West African Economic Development*, New York.

PREST, A. R. (1969). *Transport Economics in Developing Countries*, London.

STANLEY, W. R. (1971). Evaluating construction priorities of farm-to-market roads in developing countries: a case study, *The Journal of Developing Areas*, **5**, 371–400.

TAAFFE, E. J. *et al.* (1963). Transport expansion in underdeveloped countries: a comparative analysis, *Geographical Review*, **53**, 503–29.

TINGLE, E. D. (1971). Technical aid for developing countries, *Roads and Road Construction*, **49** (585), 308–13.

WARD, MARION W. (1969). Progress in transport geography, in *Trends in Geography*, (eds. R. U. Cooke and J. H. Johnson), Oxford, pp. 164–72.

WILSON, G. W. *et al.* (1966). *The Impact of Highway Investment on Development*, Washington.

VOIGT, F. (1967). *The Importance of the Transport System for Economic Development Processes*, Addis Ababa, United Nations Economic Commission for Africa, E/CN. 14/CAP/39.

1 Geography, Transportation and Regional Development

HOWARD L. GAUTHIER

SINCE World War II no single problem has commanded the attention of social scientists more than that of economic development. Efforts to understand the process of modernisation and to find effective techniques for accelerating economic growth have generated an enormous literature in economics, political science, sociology and planning. Incredibly, American geography has remained outside of this mainstream of activity.

In a recent survey, Keeble found that of the more than 500 major articles which have appeared in *Economic Geography* and the *Annals* between 1955 and 1964 only 16 were concerned in whole or part with questions related to economic development. Moreover, the treatment of development issues showed a focus quite different from that which has been characteristic of the other social sciences. Notably lacking was an analytical approach to problem solving. Four of the articles were descriptive accounts of the relationship between the physical environment and economic development, three were classifications of areas in terms of various indices of economic development, and the largest group of six articles emphasised the unique characteristics of individual areas with only peripheral reference to questions of development (Keeble, 1967).

Clearly, the record of American geography in pursuing research on developmental problems is less than impressive. This is unfortunate given the recent concern that has arisen in many related disciplines over the inadequacies of the existing theories and models of development. In the field of economics, for example, analysis has focused traditionally on the need to raise the proportion of the Gross National Product devoted to capital formation. This approach has emphasised sectorial commitments. Recently, there has been recognition of the limitations of considering the allocation of scarce resources without attention to the related issue of where given activities should be located. This question of where activities should be located is leading economists to a consideration of the regional component of economic development and to the construction of models of the spatial structure of the economy. This is an area of concern which lies traditionally within the core of geographic research. The work of Platt and Philbrick, among others, has emphasised a concern with the spatial perspective as evident in the ways in which geographic phenomena are spatially interconnected and the interaction that occurs over geographic space.

In treating the complex process of development in a regional context there are many divisions of geography which make direct contributions. The purpose of this paper is to consider transportation geography as one such division that is relevant to the study of the spatial structure of the economy. Specifically, this paper will review some traditional ways of viewing transportation in the development process, argue the necessity of considering transportation as a spatial system in regional development, and suggest some problems that arise in attempting to relate the spatial impact of transportation to goals of regional integration.

NONSPATIAL RELATIONSHIPS BETWEEN TRANSPORTATION AND ECONOMIC DEVELOPMENT

If there is a relationship between capital formation and economic growth, there must be a relationship between important components of capital formation and growth. Undoubtedly transportation is an important component of capital formation. In underdeveloped countries, it is generally the largest expenditure in the national budget, and in combination with other types of economic infrastructure represents a greater commitment in capital formation than that being made for social welfare. Typically, the proportion of public expenditures devoted to transportation investment ranges between 20 and 40 per cent. In addition, 20 per cent or more of developmental loans made by various United States and international lending agencies have been for investments in transportation (Fromm, 1965).

That economic development requires adequate and effective transportation services is axiomatic. However, after almost 20 years of study, there exists no consensus on the role of transportation in the development process. A review of the literature suggests three possible relationships, with transportation having: (1) a positive effect on the development process—the expansion in directly productive activities being a direct result of providing improved transportation facilities; (2) a permissive effect on the development process, because transportation does not independently produce directly productive activities or subsequent increases in the level of economic growth; (3) a negative effect occurring when an over-investment in transportation reduces potential growth in directly productive activity and, consequently, leads to an absolute decline in the level of income *per capita*.

The historic and most common view of the role of transportation in the development process is as a precondition or prerequisite for economic growth. In identifying the stimulus for the take-off stage of economic growth in the United States, Rostow identifies the railroads as the critical investment sector (Rostow, 1964). In accordance with this viewpoint, Hunter suggests that the economic history of Western Europe and North America has shown that the introduction of modern transportation methods has drastically lowered shipping costs. The effect has been to widen markets and to permit economies of large-scale production in a wide range of activities. According to Hunter, there is a causal linkage between low-cost transportation and economic development; the industrial revolution was successful because of a priori revolution in transport technology (Hunter, 1965). It is not surprising, in view of the historical importance assigned to transportation in the development process, that students of economic development today should look to the transportation sector as a critical component. Owens, for example, suggests that a good case can be made for transportation as the key to national development on the grounds that the widening of domestic markets is essential to economic growth (Owen, 1964).

In recent years there has been a movement away from the acceptance of transportation as the causal factor in economic development. While recognising its importance, there is more emphasis on its permissive role. This viewpoint argues that development is not a deterministic process and the singling out of a single component of capital formation as a causal agent is a gross oversimplification of a very complex problem. Hirschman (1958*a*) adopts this view in considering transportation and its relationship to directly productive activities in terms of a sequence of induced decision-making processes. Implied is the idea that the development process involves a complex interaction between human and material resources with investment in transportation offering the possibility for developing other resources. As Hawkins has observed, transportation improvements may release working capital which can be used more productively as fixed capital elsewhere, but before any of this takes place, there must be suitable productive opportunities in potential markets (Hawkins, 1962).

The permissive view of transportation has led to considerable empirical work aimed at a re-examination of American transport history. In particular, the role of the railway in the North American experience has received considerable attention. Contrary to the view that the railroads were a prerequisite for economic growth, Cootner argues that railroad growth following 1830 did not precede the growth of the other sectors of the economy but rather followed them (Cootner, 1963). A similar work by Fogel (1964), shows that the railroad was not indispensable to American growth and that in fact transportation capacity in the United States, until the turn of the century, could have been provided by existing waterways. In terms of a causal association, the railroads were built to demand and not in advance of demand.

As a result of the critical re-examination of the role of transportation in the development process, a third viewpoint has emerged which argues that transportation may have a negative impact on economic growth. Essentially this viewpoint contends that the creation of transportation capacity may absorb some portion of scarce resources that should be employed elsewhere. Specifically, in terms of opportunity costs, the investment is considered less productive than some alternatives, and as a consequence, keeps the growth rate below what it would be if those resources were used more efficiently. Basically it is a case of misdirected investment.

Errors in the allocation of resources can occur in any sector of the economy. Errors are inevitable in the sense that some other allocation could have yielded better results, although this cannot always be foreseen. Many economists argue that this is especially likely to happen in the transportation sector for two reasons.

(1) The lumpiness, longevity, and externalities associated with transportation capital create greater hazards in calculating and specifying future benefits and costs. This makes decisions to invest in transportation not easily reversible nor as readily corrected as in those sectors with assets that wear out rapidly or can be built in small increments (Wilson *et al.*, 1966).

(2) There is a belief that transport is a safe investment politically. Hirschman suggests that perhaps it is the absence of criteria and of sanctions that so much endears transportation investments to developers. After all, development planning is a risky business and there is naturally an attraction to undertake ventures that cannot be proven wrong before they are started and are unlikely ever to become obvious failures (Hirschman, 1958*b*).

TRANSPORTATION AND UNBALANCED ECONOMIC DEVELOPMENT

The recognition of the overemphasis of transportation as a causal mechanism in development has focused attention on the general relationship between social overhead capital and direct productive activities. Obviously some social overhead capital investment is

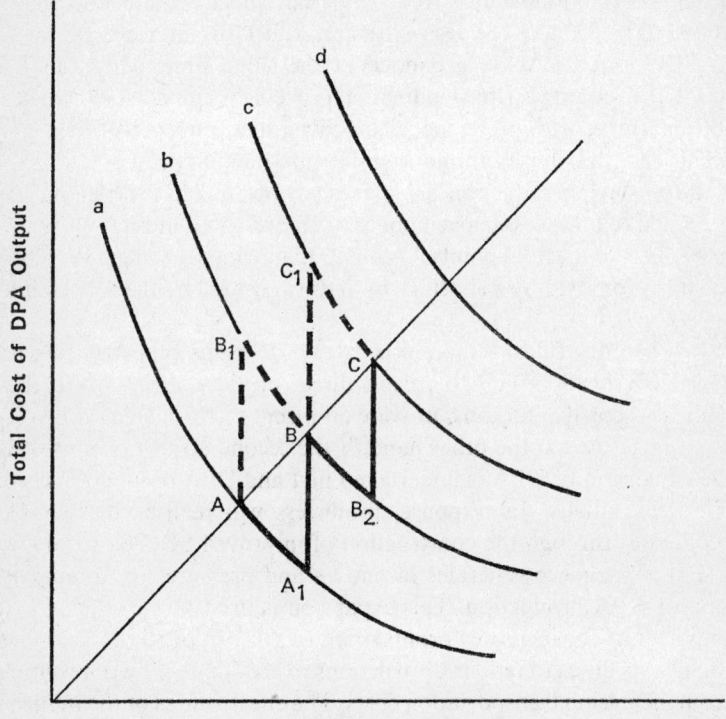

SOC Availability and Cost

Fig. 1.1   Balanced and unbalanced growth of DPA and SOC (after Hirschman, p. 87)

required as a prerequisite for direct productive activity, but within rather wide limits the relationship between the two is not technologically determined. As Hirschman has observed, it is conceivable that the relationship can be balanced or unbalanced through time (Hirschman, 1958c). The possibilities are illustrated in Fig. 1.1.

The total cost of DPA output is measured on the vertical axis, while the availability and cost of social overhead capital is indicated on the horizontal axis. The curves *a* through *d* represent successively higher amounts of DPA output. From the viewpoint of the economy as a whole, the goal is to obtain increasing outputs of DPA at minimum cost in terms of the resources devoted to both DPA and SOC.

Theoretically this should result in the most economical utilisation of a nation's resources, but as Hirschman observes, one of the paradoxes of development is that underdeveloped countries cannot afford to be economical. It is extremely difficult to allocate resources so SOC and DPA are expanded at the same time. Thus, underdeveloped countries must pursue a process of unbalanced growth through time, with preference being given to a sequence of developments that maximises an induced decision-making process.

Two principal sequences are suggested: one in which development is related to an excess capacity of SOC (represented in Fig. 1.1 by the heavy line connecting points AA_1BB_2C); and another where development is promoted by a shortage of SOC (indicated by the dashed line AB_1BC_1C).

If a country follows the first strategy, it begins by expanding its social overhead capital to permit direct productive activity to become less costly, and thus, provide an incentive for increased investment in DPA. On the other hand, if the second strategy is pursued, the expansion of DPA is undertaken first and DPA production costs rise substantially. In response, producers will realise considerable economies through the construction of improved SOC facilities.

Either sequence generates incentives and pressures for an expansion of DPA production. This expansion is in response either to an opportunity for increased profits or an increase in public expenditure in SOC facilities to reduce obstructions to economic growth resulting from increasing transportation costs. The effectiveness of the induced decision-making process will depend on profit motivation and on the response of the authorities responsible for investment in SOC to public pressure.

TRANSPORTATION AS A SPATIAL SYSTEM IN THE DEVELOPMENT PROCESS

The argument that economic growth should be viewed sectorally as an unbalanced process raises the companion problem that it be viewed as an unbalanced process in geographic space. Perroux in his well-known article on growth poles argues that a fundamental fact of sectorial development is that growth does not appear everywhere nor simultaneously. Rather it appears at points or development poles with varying intensities and spreads along diverse channels with varying terminal effects for the economy (Perroux, 1964). Hirschman argues that for an economy to attain higher income levels it must develop several regional centres of economic strength (Hirschman, 1958e). The presence of 'growth poles' in the process of economic development means that interregional inequality of growth is an inevitable concomitant and condition of growth itself.

In analysing economic growth as an unbalanced process both sectorally and spatially, many of the traditional models dealing with spatial variation in levels of development are irrelevant. Most of our interregional growth models are based on concepts drawn from international trade theory. As a consequence they are dependent on static equilibrium and assume that, given the relatively free mobility of the factors of production, factor movements tend to bring about an equalisation of income among regions. As Slater has noted, such equalisation models are of little use in illuminating the development of spatial variation in the real world, since regional inequality is not only remarkably persistent but apparently increasing in many countries (Slater, 1968).

Intuitively appealing in treating regional inequalities is the notion of growth poles. As conceived by Perroux, growth poles develop in an economic space which is defined without reference to geographic space. The distinction between the economic space in which growth poles are defined and the geographic space in which they happen to have a location is a basic and important one which has been neglected too often by those using the concepts of growth poles. The growth pole concepts *a priori* do not offer any explanation of the location of a propulsive industry in geographic space nor the consequences of a pole having a location in a given geographic space.

Much of the French economic literature during the 1950s developed extensions of the growth pole notion without reference to geographic space. Most of those studies attempted to examine inter-industry linkages, to rank industries by their degree of independence, and to show that some sectors have a very high combined linkage impact, both forward and backward, and presumably exert a polarising influence on the spatial economy.

Focusing on interindustry linkages and ignoring questions of the spatial incidence of growth is one of the great shortcomings of the original growth pole idea. As Darwent has observed, 'since all economic units must have a location, and since in regional economic development the question of "where?" looms large, then despite the fact that poles are independent of geographic space their existence within it poses complex problems unexplained by growth pole theory' (Darwent, 1968a). To meet this inadequacy the original growth pole concepts have been broadened to include geographic space. In contrast to Perroux' non-geographical orientation is Boudeville's emphasis on the regional character of economic space (Boudeville, 1961). He maintains that from a development viewpoint there are three types of geographic space: homogeneous, polarised and programmed or planning space.

Homogeneous geographic space is equivalent to the uniform region and is characterised by a maximum internal homogeneity for whatever phenomena is being measured. Polarised space is very similar to Robert Platt's concept of the functional region; the emphasis being on the linkages that exist between points distributed in geographic space and the intensity of interaction associated with those linkages. As such, polarised space is compatible with the central place structure of a hierarchy of cities of ascending size and function. The propulsive industries that create economic growth poles have a geographic location in growth centres which are the larger, more functionally complex centres in the urban hierarchy.

Finally, a region can be defined from the point of view of specific planning goals. A planning or programming region is geographic space organised for the realisation of the objectives of a planning or political authority. The concept of a planning region has taken on special significance in France, where the system of national and regional economic planning calls for the definition of regions and advice on the spatial as well as sectorial distribution of investments.

A vital question in programming the spatial incidence of economic growth is the regional impact of transportation investment. What degree of interdependence exists between the development of a transportation system and a geographic pattern of urban economic growth?

One may consider capital investments that lead to additions or changes in the transportation network as shocks that are felt throughout the entire system. One possible consequence of those shocks is an alteration in the spatial structure of the network. The change in network structure has an impact on economic development by changing the pattern of internal accessibility for urban centres on the network.

Changes in the accessibility for a set of urban centres threatens to disrupt the existing patterns of spatial competition within the region. This in turn may have a decided impact on relative rates of urban growth (Gauthier, 1968*b*).

Just as transportation investment may have positive or negative sectorial consequences, it also may have positive or negative spatial consequences. Certain centres are advantaged by their increase in accessibility while others are disadvantaged. A changing pattern of accessibility means a change in incidence of growth. This, in turn, has ramifications for programming space in that a changing pattern of accessibility poses the problem of determining whether or not the changes in the spatial structure of the economy are those desired. Are they consistent with the objectives of a given regional plan for development?

It may be that the heavy investment many underdeveloped countries are making in transportation is creating a polarised space that is inconsistent with the spatial objectives of their regional development programmes. For example, in developing its programme of regional development, the planning agency of the State Government of São Paulo, Brazil, has sought to reduce regional differences in economic well-being by influencing the spatial incidence of economic growth at designated regional centres. These centres are to serve as secondary targets for the factor movements of labour and capital that have been attracted traditionally to the major metropolitan areas, for example São Paulo and Rio de Janeiro. The important feature of this programme, in terms of net investment, has been the improvement of transportation facilities between the proposed regional centres and the major metropolitan areas. The principal goal is to increase commercial flows between the centres by reducing average transportation costs. Theoretically, the improvements should help to create conditions that increase the attractiveness of the regional centres as foci for capital investment, given the permissive role of transportation in the process of economic development. It may be this is an unwarranted expectation (Gauthier, 1968*a*). There is no reason to assume that the improvement in accessibility of the regional centres will necessarily accelerate their rate of growth more than that of the metropolitan regions. Indeed, just the reverse may be the case. The reduction in the transport cost barrier may increase the agglomerative advantages of the metropolitan areas. As a consequence externalities will make locations in São Paulo and Rio even more attractive to propulsive industries. If this occurs, then, the changes in polarised space are inconsistent with the planning objectives of the state government.

In treating the problem of increasing polarity or primacy in the spatial incidence of growth, the work of Friedmann (1966) is intuitively appealing. Friedmann considers regional organisation in terms of a centre-periphery model. He observes that a planner might expect that capital and labour would initially combine at a small number of growth poles having high marginal productivity, but would gradually diffuse from them to other centres as the original development opportunities at the centre are exhausted, diminishing marginal returns set in, and the demand for raw materials and intermediate products indicates potentially profitable investments on the periphery. Under these conditions, one can reasonably expect that capital will tend to flow from low productivity regions to the incipient growth poles of the economy, and labour from low to high-wage areas until, by a process of successive marginal adjustments, a spatial equilibrium is established. In short, there should be a gradual convergence in the rates of return to the different factors of production employed at each location. The principal difficulty with this strategy is that historical evidence does not support it. As Hirschman has argued, disequilibrium is built into transitional economies from the start, and the indisputable fact is that regional convergence will not automatically occur in the course of a nation's development (Hirschman, 1958*f*).

Friedmann suggests that the failure of convergence to occur is related to a number of factors, including: (1) a failure of diminishing returns to set in at the centre; (2) a failure to perceive investment opportunities in the periphery; (3) a growing export demand for goods produced in the centre; (4) the growing coincidence of the centre with the national market; (5) the location of quaternary services in the centre; and (6) the inability of the periphery to make adjustments appropriate to the socio-economic changes occurring at the centre (Friedmann, 1963).

The allocation of investments in geographic space involves questions of growth versus welfare, imbalance versus balance, and concentration versus dispersion. In many respects these three dichotomies are similar to each other in raising the old issue of whether to emphasise growth at the expense of welfare by concentrating investments in large agglomerations which will produce multiplier effects rather than searching for equity or balance. The choice in sectorial terms between balance and imbalance is expressed spatially as either dispersion or concentration.

Friedmann argues that the goals of a society in a transitional phase of its economy must be related to the removal of the periphery by substituting for it a single, interdependent system of urban regions and the extension of a national system of factor and commodity markets. The goals for spatial organisation and the methods of implementation require that they be related to the regional system as a whole and be consistent with dominant regional aspirations. Regrettably Friedmann does not provide us with an operational model for obtaining those goals. However, it is apparent that he views economic growth as occurring in a matrix of location points which are the building blocks around which economic space is constructed, and which evolve in the direction of ever greater spatial integration. In effect, Friedmann generalises his location points as cities and towns which serve as basic decision points in the growth process. Because of their valuable urbanisation and localisation economies, they are attractive as centres of growth. In this respect Friedmann's viewpoint is consistent with the emergence of a hierarchical system of cities. He implies that economic development is related to the emergence of a hierarchy of cities of the Lösch type with the rate of growth being some function of size, modified by imperfect labour mobility.

Obviously there are many areas in which growth pole and growth centre notions fall short of our expectations for a theory of the spatial incidence of economic growth. As Darwent has observed, the most serious omissions concern the absence of explicit statements about the relationship between polarisation and empirically observed regularities, and the inadequate treatment of the question of external economies (Darwent, 1968b). To this we might add an inadequate treatment of the role of transportation development in the organisation of the space economy. There are many important questions which need to be answered. For example, Berry (1964) stresses the relationship between the development of a central place system and a state of entropy in a socio-economic system, achieved in the steady state of a stochastic process. This seems compatible with Friedmann's idea that the spatial objective of economic development is the progressive replacement of a centre-periphery structure with a single system of cities extending throughout the economic space under consideration. However, it may be that transportation investment strengthens the centre-periphery structure of the economy rather than generating a movement toward the spatial integration envisioned by Friedmann and Berry.

Is the tendency toward spatial polarity or primacy a normal aspect of the early stages of regional development? If so, is it corrected in the process of development by the evolution of a spatial system characterised by an integrated central place hierarchy? What is the mechanism by which the evolutionary process begins and becomes self-sustaining? Does the development of linkages between propulsive industries encourage the development of a central place system which will integrate the space economy? Is there an optimal central place structure at any given level of development in the regional economy? To what extent is such a structure related to investments in regional infrastructure, particularly transportation?

At the present time all these questions are begged by the conceptions we have of regional differences in economic growth and the methods for implementing a spatial incidence of growth. Certainly they are fundamental questions we must answer if we are to understand the regional structure of the space economy and the process by which it develops. Clearly they represent problems which lie within geography's traditional concern with regional organisation and regional development. One can only hope we will begin to probe such problems in order to make a contribution toward understanding the spatial dimensions of economic development.

REFERENCES

BERRY, B. (1964). Cities as systems within systems of cities, (eds. Alonso and Friedmann), *Regional Development and Planning*, Cambridge, Mass., pp. 116–37.

BOUDEVILLE, J-R. (1961). *Les Espaces Economiques*, Paris, pp. 8–16.

COOTNER, P. (1963). The role of the railroads in U.S. economic growth, *Journal of Economic History* (December), p. 72.

DARWENT, D. (1968a). Growth pole and growth pole concepts, Working Paper No. 89, Institute of Urban and Regional Development, University of California, Berkeley (October), pp. 14–15.

DARWENT, D, (1968b). Ibid., pp. 39–43.

FOGEL, R. W. (1964). *Railroads and American Economic Growth, in Econometric History*, Baltimore.

FRIEDMANN, J. (1963). Economic policy for developing areas, *Papers, Regional Science Association*, 11, 50–1

FRIEDMANN, J. (1966). *Regional Development and Policy: a Case Study of Venezuela*, Cambridge, Mass.

FROMM, G. (ed.) (1965). *Transport Investment and Economic Development*, Washington, p. 226.

GAUTHIER, H. L. (1968a). Least cost flows in a capacitated network: a Brazilian example, *Geographic Studies of Urban Transportation and Network Analysis*, North-Western University Studies of Geography, **16**, 102–27.

GAUTHIER, H.L., (1968b). Transportation and the growth of the São Paulo economy, *Journal of Regional Science*, **8**, 77–94. See page 167 of this volume.

HAWKINS, E. K. (1962). *Roads and Road Transport in an Underdeveloped Country: a Case Study of Uganda*, H.M.S.O.: London, p. 35

HIRSCHMAN, A. O. (1958a). *Strategy of Economic Development*, New Haven.

HIRSCHMAN, A. O. (1958b). Ibid., pp. 84–5.

HIRSCHMAN, A. O. (1958c). Ibid., pp. 183–4.

HIRSCHMAN, A. O. (1958d). Ibid., p. 87.

HIRSCHMAN, A. O. (1958e). Ibid., pp. 183–4.

HIRSCHMAN, A. O. (1958f). Ibid., pp. 62–70.

HUNTER, H. (1965). Transport in Soviet and Chinese development, *Economic Development and Cultural Change*, **14**, 71–2.

KEEBLE, D. E. (1967). Models of economic development, (eds. Chorley and Haggett). *Models in Geography*, London, pp. 243–4.

OWEN, W. (1964). *Strategy for Mobility*, Washington, p. 2.

PERROUX, F. (1964). 'La notion de pole de croissance', *L'economie du XXième siècle*. 2nd edn., Paris.

SLATER, D. (1968). The modernisation process-spatial aspects and the Latin American case, Discussion Paper No. 21, London School of Economics and Political Science (July), pp. 6–10.

ROSTOW, W. W. (1964). *The Stages of Economic Growth*, Cambridge, p. 24.

WILSON, G. *et al.* (1966). *The Impact of Highway Investment on Development*, Washington, pp. 9–10.

2 Transport Expansion in Underdeveloped Countries: A Comparative Analysis

EDWARD J. TAAFFE,
RICHARD L. MORRILL AND
PETER R. GOULD†

IN the economic growth of underdeveloped countries a critical factor has been the improvement of internal accessibility through the expansion of a transportation network. This expansion is from its beginning at once a continuous process of spatial diffusion and an irregular or sporadic process influenced by many specific economic, social or political forces. In the present paper both processes are examined as they have been evident in the growth of modern transportation facilities in several underdeveloped areas. Certain broad regularities underlying the spatial diffusion process are brought to light, which permit a descriptive generalisation of an ideal-typical sequence of transportation development. The relationship between transportation and population is discussed and is used as the basis for examination of such additional factors as the physical environment, rail competition, intermediate location and commercialisation. Throughout the study, Ghana and Nigeria are used as examples. (Other areas examined in some detail, though from secondary sources, are Brazil, Kenya, Tanganyika and Malaya.) This study is based on a combination of findings in Taaffe and Morrill, 1960; Gould, 1960.

SEQUENCE OF TRANSPORTATION DEVELOPMENT

Figure 2.1 presents the authors' interpretation of an ideal-typical sequence of transport development. The first phase (A) consists of a scattering of small ports and trading posts along the seacoast. There is little lateral interconnection except for small indigenous fishing craft and irregularly scheduled trading vessels and each port has an

---

† This is a shortened version of the study published by Professor Taaffe and his colleagues in 1963.

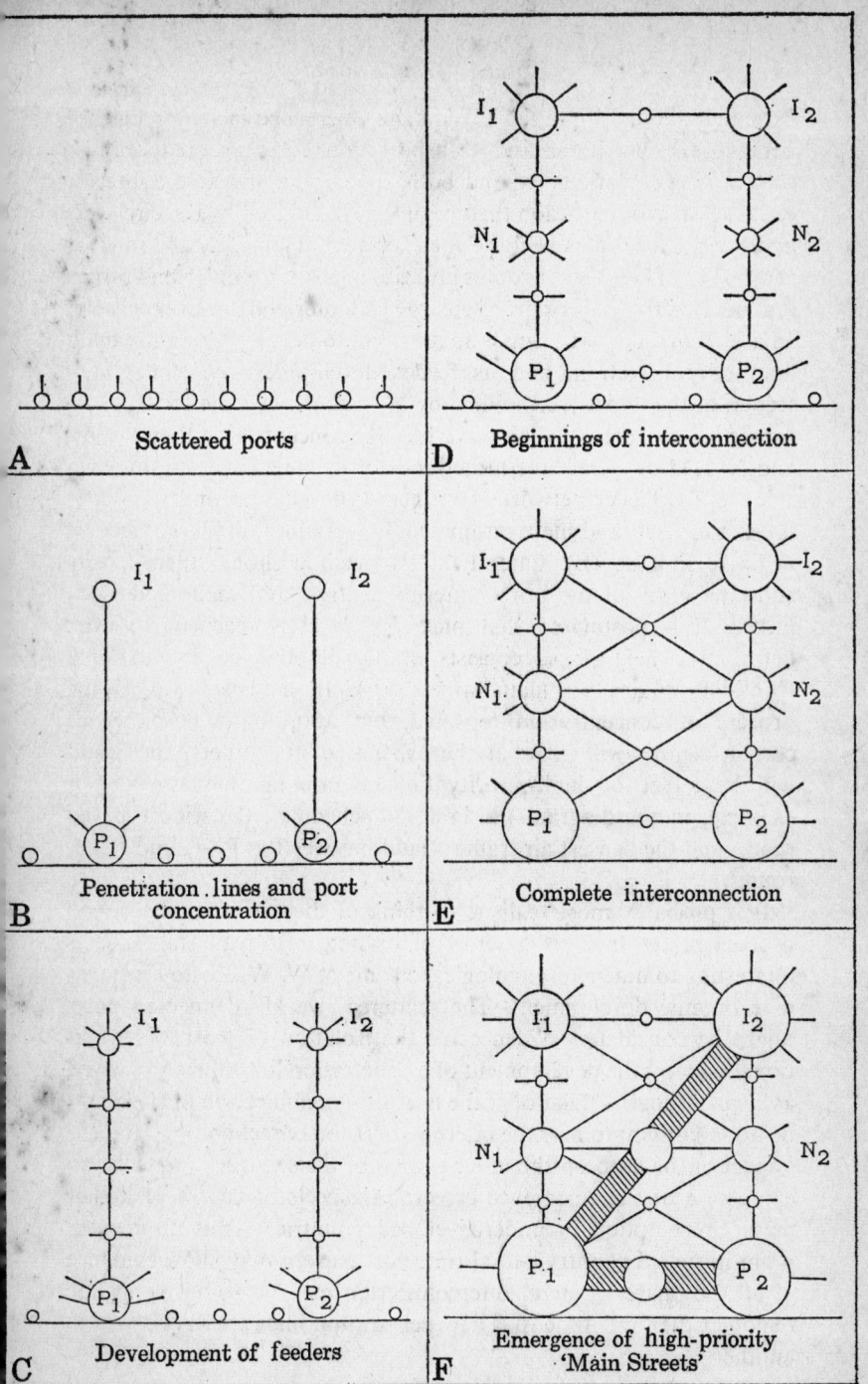

A Scattered ports

B Penetration lines and port concentration

C Development of feeders

D Beginnings of interconnection

E Complete interconnection

F Emergence of high-priority 'Main Streets'

Fig. 2.1 Ideal-typical sequence of transport development

extremely limited hinterland. With the emergence of major lines of penetration (B), hinterland transportation costs are reduced for certain ports. Markets expand both at the port and at the interior centre. Port concentration then begins, as illustrated by the circles P_1 and P_2. Feeder routes begin to focus on the major ports and interior centres (C). These feeder routes give rise to a sort of hinterland piracy that permits the major port to enlarge its hinterland at the expense of adjacent smaller ports. Small nodes begin to develop along the main lines of penetration, and as feeder development continues (D), certain of the nodes, exemplified by N_1 and N_2, become focal points for feeder networks of their own. Interior concentration then begins, and N_1 and N_2 pirate the hinterlands of the smaller nodes on each side. As the feeder networks continue to develop around the ports, interior centres, and main on-line nodes, certain of the larger feeders begin to link up (E). Lateral interconnection should theoretically continue until all the ports, interior centres, and main nodes are linked. It is postulated that once this level is reached, or even before, the next phase consists of the development of national trunk-line routes or 'main streets' (F). In a sense, this is the process of concentration repeated, but at a higher level. Since certain centres will grow at the expense of the others, the result will be a set of high-priority linkages among the largest. For example, in the diagram the best rail schedules, the widest paved roads, and the densest air traffic would be over the P_1–I_2 and P_1–P_2 routes.

It is probably most realistic to think of the entire sequence as a process rather than as a series of discrete historical stages. It is interesting to note a few analogies to some of W. W. Rostow's stages of economic development. The scattered, weakly connected ports might be considered evidences of the isolation of Rostow's traditional society; the development of a penetration line might be viewed as a sort of spatial 'take-off'; the lateral-interconnection phase might be a spatial symptom of the internal diffusion of technology; and the impact of the auto on the latter phases of the sequence might be an expression of the emergence of certain aspects of an era of higher mass consumption in underdeveloped countries. Thus at a given point in time a country's total transport pattern may show evidence of all the phases. Lateral interconnection may be going on in one region at the same time that new penetration lines are developing in another.

THE FIRST PHASE: SCATTERED PORTS

In both Ghana and Nigeria† an early period of numerous small, scattered ports and coastal settlements with trading functions may be easily identified (Figs. 2.3 and 2.4). These settlements, most of which existed or came into being between the end of the fifteenth century and the end of the nineteenth, were populated by the indigenous people around a European trading station or fort. Many of the people engaged in trade with the Europeans and served as middlemen for trade with the interior, a function jealously guarded for centuries against European encroachment. Penetration lines to the interior were weakly developed, but networks of circuitous bush trails connected the small centres to their restricted hinterlands. River mouths were important, particularly in the Niger delta, but with a few exceptions during the early periods of European encroachment the rivers did not develop as the main lines of thrust when penetration began. Most of these early trading centres have long since disappeared, destroyed by the growth of the main ports, or else they linger on as relict ports, with visits of occasional tramp steamers to remind them of their former trading heyday.

THE SECOND PHASE: PENETRATION LINES AND PORT CONCENTRATION

Perhaps the most important single phase in the transportation history of an underdeveloped country is the emergence of the first major penetration line from the seacoast to the interior. Later phases typically evolve around the penetration lines, and ultimately there is a strong tendency for them to serve as the trunk-line routes for more highly developed transportation networks. Three principal motives for building lines of penetration have been active in the past (1) the desire to connect an administrative centre on the seacoast with an interior area for political and military control; (2) the desire to reach

† For Ghana, the examples are based on the field data and primary statistical source material gathered by P. R. Gould. For Nigeria, primary statistical sources are indicated where necessary. Secondary sources include K. O. Dike: *Trade and Politics in the Niger Delta, 1830–1885*, Oxford Studies in African Affairs, London 1956; G. Walker: 'Traffic and Transport in Nigeria: The Example of an Underdeveloped Tropical Territory', *Colonial Research Studies No. 27*, Colonial Office, London 1959; *The Economic Development of Nigeria*, International Bank for Reconstruction and Development, Baltimore 1955. The authors are also indebted to Dr Akin Mabogunje, of the University, Ibadan, for his comments.

areas of mineral exploitation; (3) the desire to reach areas of potential agricultural export production. In the cases examined, the political motive has been the strongest. Political and military control dominated official thinking of the day in Africa, often as a direct result of extra-African rivalries. The second motive, mineral exploitation, is typically associated with rail penetration. It is today probably the principal motive for the building of railways in Africa, and then only after careful surveys and international agreements have virtually guaranteed the steady haul of a bulk commodity to amortise the loans required for construction. For example, the extension of the Uganda railway to Kasese to haul copper ore; the long northward extension of the Cameroon railway from Yaoundé to Garoua to haul manganese; and the new railway from Nouadhibou to F'Derik in Mauritania to haul iron ore.

The development of a penetration line sets in motion a series of spatial processes and readjustments as the comparative locational advantages of all centres shift. Concentration of port activity is particularly important, and the ports at the termini of the earliest penetration lines are usually the ones that thrive at the expense of their neighbours (Fig. 2.2). Typically, one or two ports in a country dominate both import and export traffic, and often the smaller ports have lost their functions in external commerce.

In Ghana several interesting variations on the penetration theme appear (Fig. 2.3). The desire to reach Kumasi, capital of a then aggressive Ashanti, formed the essentially military-political motive for the first penetration road, which followed an old bush track, sporadically cleared whenever the local people were goaded into activity. The road was built from Cape Coast, and although it is still important as one of Ghana's main north–south links, the port function of Cape Coast declined as Sekondi increased in importance. Sekondi's great impetus came with the building of the rail penetration line to Kumasi at the turn of the century, after which adjacent ports such as Axim, Dixcove, Adjua, Shama, Komena and Elmina suffered a rapid decline in traffic. The Pra and Ankobra Rivers, east and west of Sekondi, which were formerly of some significance as avenues of penetration, also experienced marked traffic decreases. The initial motive for the western railroad was primarily mineral production (the goldfield at Tarkwa) and secondarily provision of a rapid connection for administration between the seacoast and a troublesome internal centre of population.

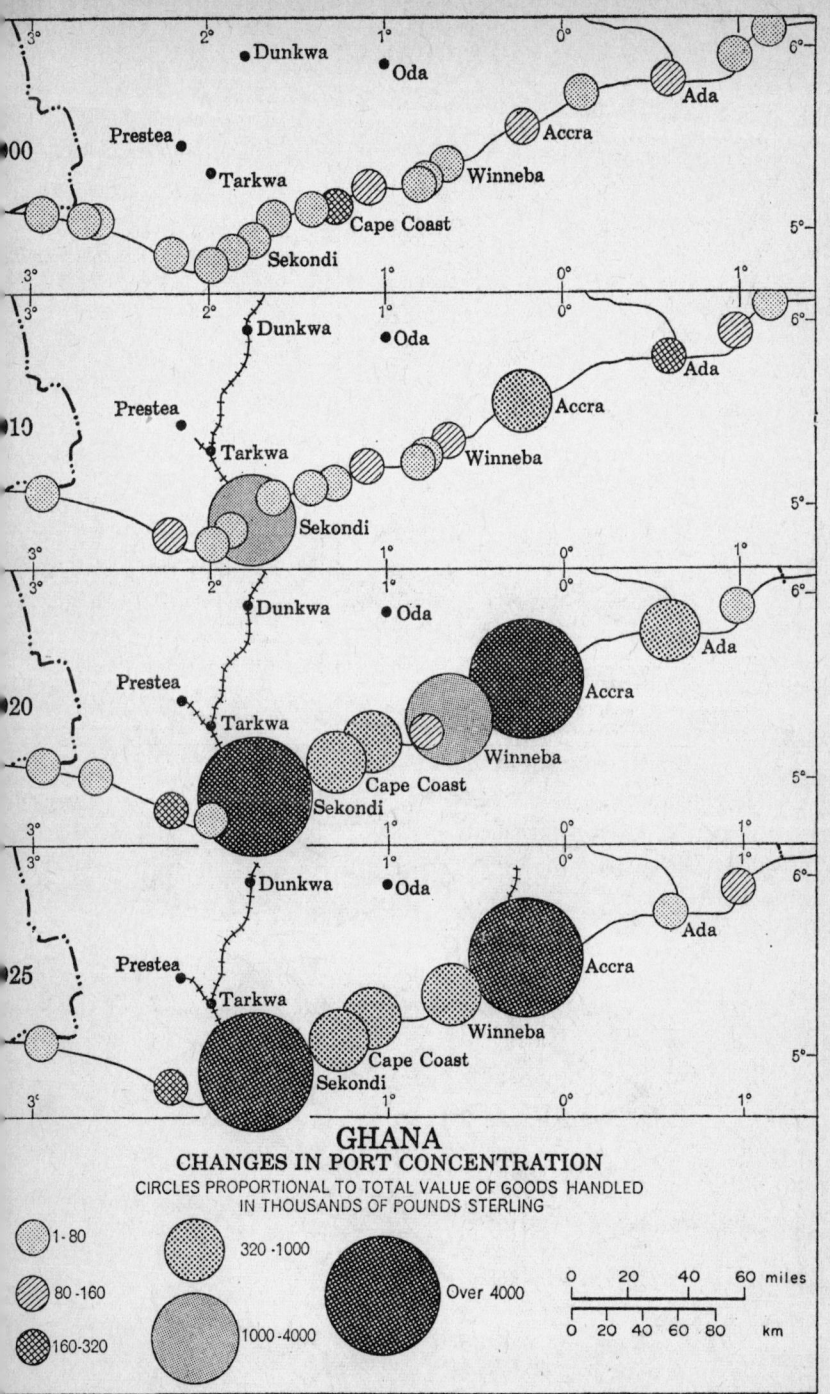

Fig. 2.2 *Changes in port concentration, Ghana, 1900–25 (after Gould, ibid., p. 45)*

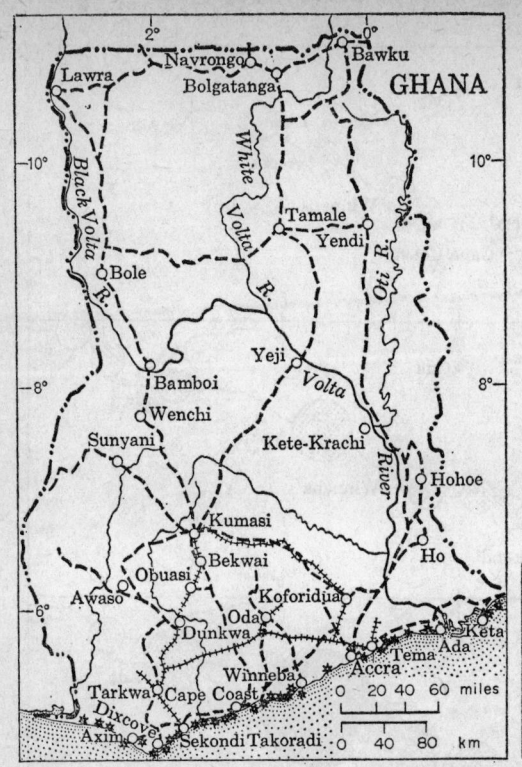

Fig. 2.3 Major transport facilities Ghana (after Gould, ibid., p. 45)

MAJOR TRANSPORT FACILITIES

+++++++++ Railroads
--------- Main roads
* Former trading posts

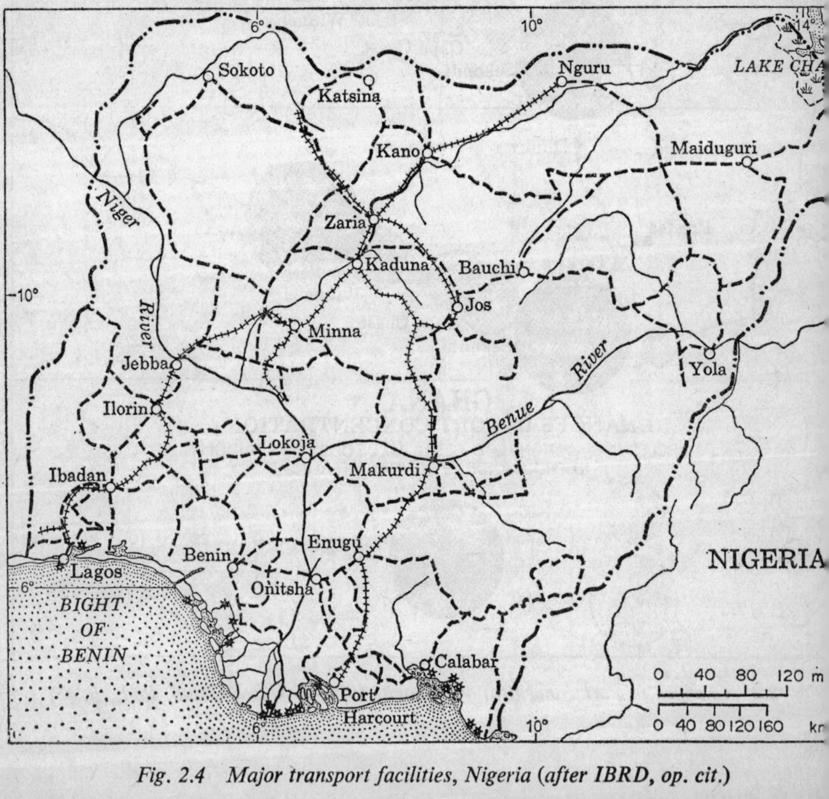

Fig. 2.4 Major transport facilities, Nigeria (after IBRD, op. cit.)

The eastern railroad penetration line was slower in developing, partly as a result of the interruption of the First World War and of smallpox outbreaks in the railroad camps. The link between Kumasi and Accra was not completed until 1923, twenty years after the Sekondi–Kumasi link. Connection with the rapidly expanding cocoa areas north of Accra was the immediate reason for this line, underlain by the political desire to connect the leading city, Accra, with Kumasi, the main population and distribution centre in the interior. As in the case of Sekondi–Takoradi, Accra's importance increased steadily at the expense of adjacent ports as the railroad penetrated inland.

The two penetration lines forming the sides of the rail triangle were now complete, and a considerable amount of subsequent transportation development of the country was based on these two trunk lines. Penetration north of Kumasi was entirely by road, despite grand railroad plans at one time. There were no minerals to provide an economic incentive for railroads, and the barren middle zone, which separated the rail-triangle area from the densely settled north, acted as a deterrent to the continuation of the railroad in short stages.

The Great North Road is the chief line of penetration north of Kumasi, built with a strong political-administrative motive to Tamale, a town deliberately laid out as the capital of the Northern Territories in the early years of this century. Early feeder development focusing on Tamale helped to fix the position of this trunk line and its extension to the northern markets of Bolgatanga, Navrongo and Bawku. The Western Trunk Road was built with similar motives and grew out of the extensive Kumasi feeder network to link that city with a moderately populated area north of the barren middle zone. In the building of the Eastern Trunk Road political and economic motives were mixed. This road through former British Togoland was originally extended from the Hohoe cocoa area to Yendi for transport of yams to the rapidly growing urban centres of the south, but its extension to the northern border hinged in large part on political motives that were very strong immediately before the United Nations plebiscite and Togo's resultant political affiliation with Ghana. The original road links from the rail triangle and Accra to the Hohoe cocoa area were associated with a political desire to forge a link with British Togoland, and with an economic desire to prevent the diversion of this area's cocoa traffic to the port of Lomé in then French Togoland.

The process of penetration and port concentration in Nigeria (Fig. 2.4) was markedly similar to that in Ghana; the main difference lay in the greater emphasis on long-haul rail development and the subsequent higher level of economic development in the north. Again the initial motives were somewhat more political than economic; for even the early penetration to the north via the Niger River by the Royal Niger Company had imperialistic as well as economic motives. In a sense, Kano might be regarded as analogous to Kumasi. Both are important interior centres which predate European settlement and which were later connected to the main ports by rail penetration lines. The chief differences, of course, are the vastly greater distance between Kano and the coast and the greater width of Nigeria's relatively barren middle zone. Mineral exploitation was also a major motive for the building of rail penetration lines in Nigeria, particularly the eastern railroad. The line from Port Harcourt was started in 1913 and was connected with the important Enugu coalfields three years later. This port serves also as the principal outlet for the tin outputs of the Jos Plateau. The connecting of agricultural regions to the coast, though not a strong initial penetration motive, was apparently associated with the actual linking of the northern and southern lines.

As in Ghana, the rail penetration lines form the basis for the entire transportation network. The only area of extensive road penetration is in the northeast, from the railroad at Jos and Nguru to Maiduguri. The main motive for establishing tarred roads and large-scale trucking services was the attraction of the Lake Chad region to the northeast. However, the Maiduguri region has now been connected by rail to the main network, despite the recommendation of a mission of the International Bank for Reconstruction and Development, which felt that roads could more efficiently accommodate the expected increase in traffic. In the southeast there is no effective road or rail penetration line.

Port concentration has been marked. The decline of the delta ports began with the building of the rail penetration line from Lagos and was accentuated by the building of the eastern line and the concomitant growth of Port Harcourt. In 1958 these two major ports accounted for more than three-quarters of Nigeria's export and import trade.

THE THIRD PHASE: FEEDERS AND LATERAL INTERCONNECTIONS

Penetration is followed by lateral interconnection as feeder lines begin to move out both from the ports and from the nodes along the penetration lines. The process of concentration among the nodes is analogous to the process of port concentration; it results when the feeder networks of certain centres reach out and tap the hinterlands of their neighbours. As feeder networks become stronger at the interior centres and intermediate nodes some of them link and thereby interconnect the original penetration lines. The degree of interconnectedness of a transport network could be precisely evaluated by the use of such new measures as those presented in Garrison, 1961.

Figure 2.5 presents a sequence of road development in Ghana from 1922 to 1958. The shading represents road-mileage density as recorded in a series of grid cells of 283 square miles superimposed on a highway map. In 1922 Ghana had just entered the phase of lateral interconnection, with east–west linkages both in the south, along the coast, and among the centres of the north, and with an extensive feeder network steadily drawing more and more of the smaller population centres into the orbit of Kumasi. Development in the southwest was weak, owing to railroad competition and to a deliberate policy of maintaining an economic road gap between Sekondi–Takoradi and Kumasi. This gap was finally filled in 1958, and only now is the southwest beginning to realise its great potential in cocoa and timber. By 1937 lateral interconnection had become more marked. The connections east and west of Tamale provide a good example of links between intermediate nodes. The 1–20-mile shading, for example, now reaches west from the Tamale node on the Great North Road to the node at Bole, which was just developing on the Western Trunk Road in 1922; Yendi on the Eastern Trunk has been similarly linked. Lateral interconnection has become intensified in the north, and the 21–40-mile shading now covers the entire zone between Bawku and Lawra. In the Kumasi area feeder development has continued, and the 21–40-mile shading blankets the Wenchi–Sunyani area to provide a fairly good network of interconnections between the Western Trunk and the Great North Road in the zone where both converge on Kumasi. Urban geographers will note the strong analogy to the process of interstitial filling between major radial roads converging on a central business district.

In 1958 the lateral interconnection process is fairly well developed. Only a few areas are still without road links, formerly inaccessible areas having been tapped by the expanding road network. A new series of high-density nodes have developed since 1937 and already are reaching out toward one another. Marked examples occur in the

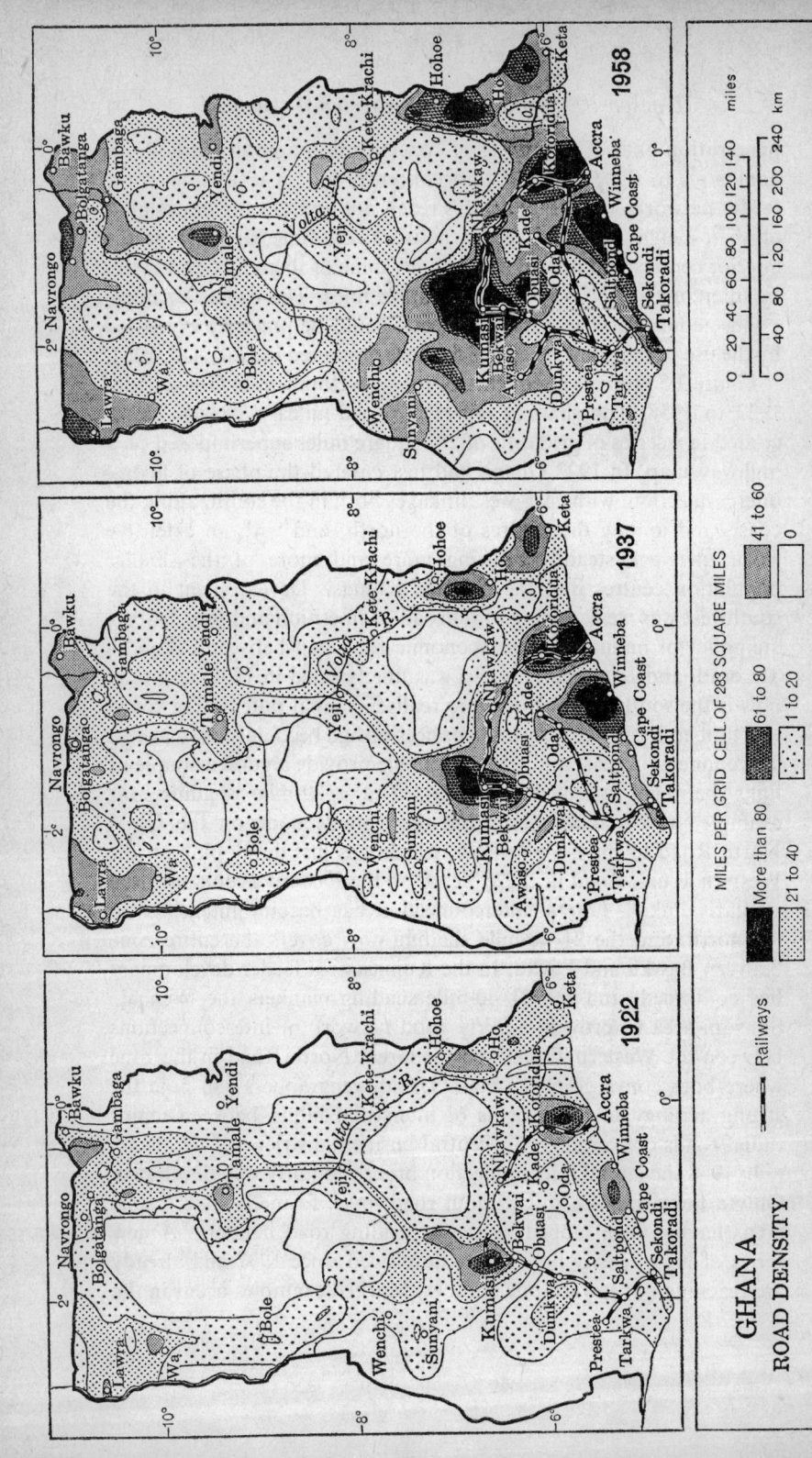

GHANA

ROAD DENSITY

MILES PER GRID CELL OF 283 SQUARE MILES

More than 80   61 to 80   41 to 60

21 to 40   1 to 20   0

Railways

0   20   40   60   80   100   120   140   miles

0   40   80   120   160   200   240   km

1922

Navrongo
Bawku
Bolgatanga
Gambaga
Lawra
Wa
Yendi
Tamale
Bole
Yeji
Kete-Krachi
Volta R.
Wenchu
Sunyani
Kumasi
Bekwai
Obuasi
Awaso
Kade
Oda
Nkawkaw
Koforidua
Hohoe
Ho
Dunkwa
Prestea
Tarkwa
Saltpond
Accra
Winneba
Cape Coast
Sekondi
Takoradi
Keta

1937

Navrongo
Bawku
Bolgatanga
Gambaga
Lawra
Wa
Tamale Yendi
Bole
Yeji
Kete-Krachi
Volta R.
Wenchu
Sunyani
Kumasi
Bekwai
Obuasi
Awaso
Kade
Oda
Nkawkaw
Koforidua
Hohoe
Ho
Dunkwa
Prestea
Tarkwa
Saltpond
Accra
Winneba
Cape Coast
Sekondi
Takoradi
Keta

1958

Navrongo
Bawku
Bolgatanga
Gambaga
Lawra
Wa
Tamale
Yendi
Bole
Yeji
Kete-Krachi
Volta R.
Wenchu
Sunyani
Kumasi
Bekwai
Obuasi
Awaso
Kade
Nkawkaw
Koforidua
Hohoe
Ho
Dunkwa
Prestea
Tarkwa
Saltpond
Accra
Winneba
Cape Coast
Sekondi
Takoradi
Keta

10°   8°   6°   2°   0°

north between Tamale, Yendi, and other northern population centres, and in the south in the developing and extending nodes around Kumasi and east of the Volta River.

It is clear from the regularity of the progression of the highway-density patterns that extrapolation of the density maps to some future date would be reasonable. In a sense the map sequence is a crude predictive device. For instance, the probability of an increase in road miles for any area between two nodes is greater than that for a comparable area elsewhere.

In Nigeria a basically similar pattern had developed by 1953 (Fig. 2.6), with many of the earlier nodes of high accessibility in the south linking laterally to form an almost continuous high-density strip, broken only by the Niger River near Onitsha. Lines of penetration linking the north and the south across the barren middle zone, a feature clearly brought out by the map, are relatively weakly developed, and the degree of lateral linkage is well below that of northern Ghana. Only the Kano and Zaria nodes, in areas of high agricultural production and at the centre of strong administrative webs, and the Jos node, at the centre of the tin and columbium mineral complex, stand out as exceptions. Heavy rail competition, which resulted in severe restraints on long-haul trucking for many years, has clearly weakened the western road penetration lines, and the similarity to southwestern Ghana is strong. Areas totally inaccessible by road are still numerous, particularly in the barren middle zone and along the periphery of the country. To the political geographer the general weakness of the linkages between the Eastern, Western and Northern Regions will be of particular interest. There is, in fact, a clear visual impression that the general pattern of accessibility by road in Nigeria in 1953 is similar to that of Ghana in 1937 —hardly surprising in view of the much larger size of Nigeria, the longer distances, and the lower per capita tax base from which the greater part of road development funds must come.

THE FOURTH PHASE: HIGH-PRIORITY LINKAGES

The phase following the development of a fairly complete and coherent network is difficult to identify, and a variety of labels might be applied to it. Certainly the most marked characteristic of the most recent phase in the cases studied is the dominance of road over railroad. A common theme throughout the evolution of the transportation

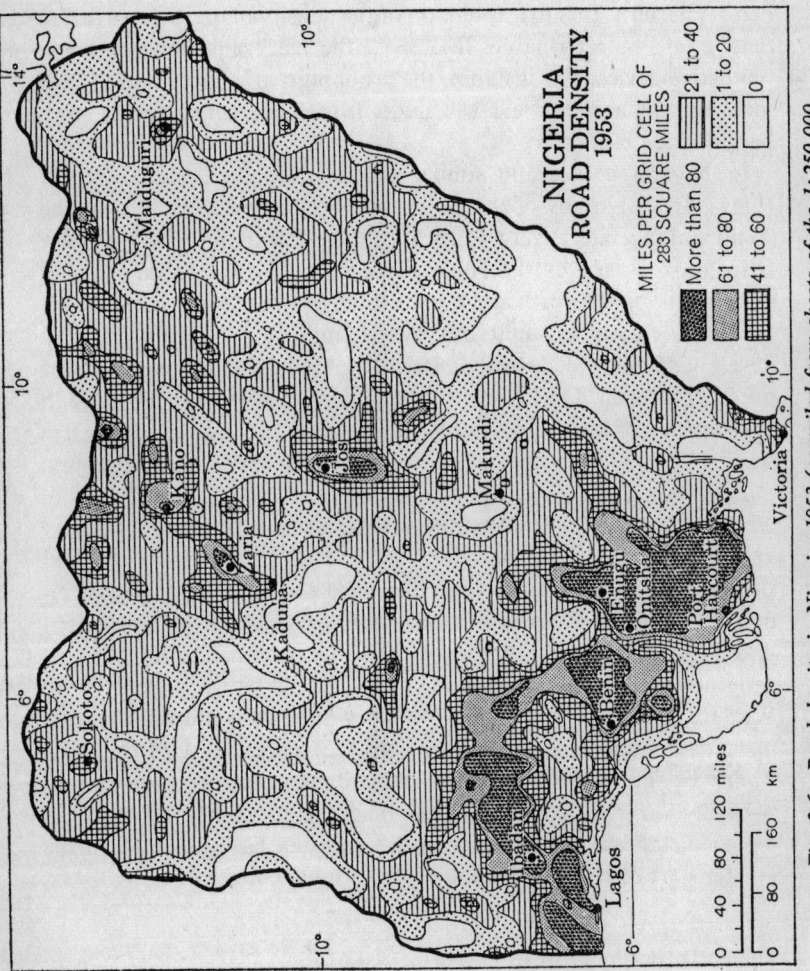

Fig. 2.6   Road density in Nigeria, 1953 (compiled from sheets of the 1:250,000 map series published by the Federal Survey Department, Lagos)

NIGERIA
ROAD DENSITY
1953

MILES PER GRID CELL OF
283 SQUARE MILES

More than 80
61 to 80
41 to 60
21 to 40
1 to 20
0

system in Ghana and Nigeria, and also in the other examples studied, has been the steady rise in the importance of road traffic, which first complements the railroad, then competes with it, and finally overwhelms it. However, the evidence available seems to indicate that this occurs irrespective of the stage of transport development, and it is possible that a greater number of road penetration lines are now being built in areas which would have required rail penetration lines in the past.

The idea of a phase of high-priority linkages is based, somewhat weakly, on a logical extrapolation of the concentration processes noted in the earlier stages of transport development in Ghana and Nigeria, and is supported in part by highly generalised evidence from areas with well-developed transportation systems.

Interior centres, intermediate nodes, and ports do not develop at precisely the same rate. As some of these centres grow more rapidly than others, their feeder networks become intensified and reach into the hinterlands of nearby centres. Ultimately certain interior centres and ports assert a geographic dominance over the entire country. This creates a disproportionately large demand for transportation between them, and since some transport facilities already exist, the new demand may take such forms as the widening of roads or the introduction of jet aircraft. In general, transport innovations are first applied to these trunk routes. For example, in the United States the best passenger rates, schedules, and equipment are usually initiated over high-density routes such as New York–Chicago. In underdeveloped countries high-priority linkages would seem to be less likely to develop along an export trunk line than along a route connecting two centres concerned in internal exchange. There is some weak evidence that high-priority links may be developing in the two study countries. High-density, short-haul traffic in the vicinity of Lagos may be the forerunner of a 'main street' between cities of the western part of the rail bifurcation. In Ghana the heavy traffic flows focusing on Accra (flows that have tripled every five years since the war) have virtually forced the authorities to bring the basic road triangle up to first-class standards of alignment and surface.

RELATIONSHIPS BETWEEN ROAD MILEAGES AND SELECTED PHENOMENA

The lateral-interconnection phase of the ideal-typical sequence is the one that best depicts the current extent of transportation development in most underdeveloped countries. This phase has been accompanied by a steady increase in the importance of motor vehicles, so that at present the dominant transport characteristic of most underdeveloped countries is the expansion of the road network. Closer examination of the road networks of Ghana and Nigeria affords deeper insight into the factors that affect the spatial diffusion of roads at a time when interconnection is a prominent motive for transport development. For example, how close a relation is there between roads and population? Do such additional factors as environment, competitive transportation, and income have an effect on the distribution of roads over and above the population effect? Attempts to answer these questions are in the form of a basic regression model supplemented by cartographic analysis. In the basic regression model, road mileage within subregional units is treated as a dependent variable, population and area as the independent variables.†

The results of the regression analysis indicate a close relationship between the internal distribution of road mileage and total population as corrected for the differing areas of reporting units. Briefly, it has been found that in a given unit, road mileage is in general proportional to the square root (approximately) of the population times the square root (approximately) of the area. Three-quarters of the internal spatial variation in road mileage is associated with these two factors alone.‡ Thus to achieve a fair first approximation of the internal distribution of road mileage at a given point in time, we look first to the population distribution. Effects of difficult terrain, unequal distribution of resources, rail competition, and the like on the distribution of roads may be regarded as being partly subsumed by the population and area variables. Much of the impact of these

---

† Sources for population figures were *Population Census of Nigeria 1952–1953*, Lagos 1954, and *The Gold Coast Census of Population, 1948: Report and Tables*, London and Accra 1950; for road mileages, *Mobil Road Map of Nigeria* 1:750,000 and 1:500,000, Federal Survey Department Lagos 1957, and *Road Map of the Gold Coast*, 1:500,000, Department of Surveys, Accra 1950. Only first- and second-class roads were included. No weighting system was applied.

‡ For Ghana the regression equation was $\log Y_c = 0.1709 + 0.6285 \log X_1 + 0.4139 \log X_2$ or $Y_c = 1.482 \, X_1^{0.6285} X_2^{0.4139}$, with Y_c the estimated highway mileage, X_1 the district population in thousands, and X_2 the district area. The r^2 or explained variation was 0.75 or 75 per cent. For Nigeria, the regression equation was $\log Y_c = -0.44771 + 0.4458 \log X_1 + 0.4823 \log X_2$ or $Y_c = (X_1^{0.4458} X_2^{0.4823})/2.799$. The explained variation in this case was 81 per cent. In both cases, the particular form of the equation was the result of normalising the data by means of log transformations.

factors on the transportation system is expressed through their relationship to the population pattern.

As expected, total population accounts for more of the variation in total road mileage than area accounts for; in both Ghana and Nigeria it accounts for about 50 per cent. The addition of area as an independent variable accounts for 20 per cent more. Obviously, there is a greater need for transportation for a given population in a large unit than in a small one. Although the demand for roads generally reflects the distribution of the population, a large, sparsely settled unit will require a large per capita road investment to be served at all. Thus the relative weights of the two independent variables, and the closeness of the correlation, are seen to be significantly affected by variations in the size of the reporting units; hence the use of simple population densities would have been deceptive in that an understatement of road-mileage expectations for large, sparsely settled units would have resulted.

On the other hand, it is not clear that meaning may be ascribed to the area variable as a separate factor. The problem of modifiable areal reporting units is, in general, a difficult one. Some aspects of the problem are discussed in Robinson, 1956; Duncan *et al.*, 1961. Internal variations in size of reporting units affect the degree of apparent correlation between variables. Even if the sizes of reporting units were uniform, different correlations and different regression equations would be obtained for different levels of areal aggregation (a grid cell of 10 square miles as opposed to 100 square miles, for instance). In this case, it is best to regard the use of an area variable as a means of including the effects of any internal variations in size of reporting units on road mileage along with any effects of variations in population. Thus it can be said that three-quarters of the variation in road mileage among the areal subunits in Ghana and Nigeria is statistically associated with their combined variations in population and size. However, if the average size of the subunits were to be significantly increased or decreased, the amount of statistically 'explained' variation would be affected.

This relatively simple regression analysis may be supplemented by an attempt to uncover further possible factors that seem to be particularly relevant to the development of road transportation. In what parts of Ghana and Nigeria does there seem to be a great deal of residual variation? Where, in other words, does the population-area equation seem to give significant overestimates or underestimates of

road mileage? Examination of the residuals maps for Ghana and
Nigeria [not reproduced here] suggests five additional factors: hostile
environment; rail competition; intermediate location; income or
degree of commercialisation; and relationship to the ideal-typical
sequence. Precise quantification of these factors did not seem to be
warranted by the data. Therefore, a subjective examination was made
of the relationship between each of the five factors and the distribution of regression residuals. In many instances the lack of data
resulted in highly generalised and arbitrary quantifications.

THE ANALYSIS IN PERSPECTIVE

As population increases in an area, the demand for transportation is
intensified; as new transport lines are built into the area, a greater
population increase is encouraged, which, in turn, calls for still more
transportation. In a sense, the models artifically separated these two
effects: the ideal-typical sequence considered transportation expansion as though it were independent of population distribution; the
regressions treated transportation as though it were caused by
population. However, the residuals maps provided intuitive evidence
of the lag-and-lead nature of transport development. One might
postulate a tendency through time for these alternate overexpansions
and deficits of the transport system to become gradually smaller until
a temporary equilibrium is reached. A transport innovation or a
sudden demand for a new penetration line, such as that occasioned
by a mineral discovery, could then reactivate the process. This suggests that a possible avenue of future investigation of transport
expansion in underdeveloped countries might be the application of a
simulation models such as the Monte Carlo technique applied by
Torsten Hägerstrand in his migration studies (Hägerstrand, 1953).
The spatial evolution of a transport and population pattern might be
simulated through time by using for each stage in the process a set of
probabilities dependent on the transport and population pattern of
the preceding stage, thus bringing the essentially stochastic nature
of transportation development into the model. The direction of the
extension of a transport line from a given point might be based on
probabilities derived from factors similar to those noted in the discussion of penetration lines and the Ghana highway-density maps.

Finally, it should be noted that the generalisations in this study are
designed to provide an initial perspective on the expansion of trans-

portation in underdeveloped countries. At the moment, it is probable that the variations from the typical sequence are of more interest than their explicit application. It is to be hoped that future studies will bring about fundamental changes in the perspective presented here, at the same or a higher level of generalisation. This may be accomplished by field investigations, by the development of more useful transportation parameters, and by the application of increasingly rigorous methods of analysis and model verification.

REFERENCES

DUNCAN, O. D. *et al.* (1961). *Statistical Geography; Problems in Analyzing Areal Date*, Glencoe, Ill.

GARRISON, W. L. (1961). Connectivity of the interstate highway system, *Papers and Proc. Regional Science Assn.*, vol. 6 (6th Annual Meeting, 1960), Philadelphia, pp. 121–37.

GOULD, P. R. (1960). The development of the transportation pattern in Ghana, *Northwestern Univ. Studies in Geogr. No. 5*, Evanston, Ill.

HÄGERSTRAND, T. (1953). Innovationsförloppet ur korologisk synpunkt, *Meddelander från Lunds Univ. Geogr. Instn., Avhandl. 25.*

ROBINSON, A. H. (1956). The necessity of weighting values in correlation analysis of areal data, *Anals Assn. of Amer. Geogrs.*, **46**, 233–6.

TAAFFE, E. J. and MORRILL, R. L. (1960). Transportation Geography Research: Part 2, Investigation of the internal spatial distribution of transportation facilities', (unpublished report to the U.S. Army Transportation facilities, Research Command under the auspices of the Transportation Center at Northwestern University, July 1).

3 Transport and Economic Growth in Developing Countries: The Case of East Africa

B. S. HOYLE

INTRODUCTION

THE emergence of a viable and efficient system of modern transportation is clearly an essential element in the growing infrastructure upon which the expanding economies of the less developed countries must be based. Many analyses of transport problems in such countries emphasise the specific problems associated with road, rail or sea transport media, and frequently confine discussion either to the economics of transportation or to the geographical disposition of the facilities available. Relatively few studies attempt to view the economic geography of transport networks within an area as a whole, or to examine the nature of the relationship between transport provision and economic growth. These two aspects of transport geography are of vital significance in the less developed countries, especially in view of the need to achieve maximum economies within the development process. East Africa provides an example of an area where rail, road, sea and air transport systems are all reasonably well developed but lack close integration, and where existing networks are currently subject to considerable changes and are, therefore, objects of political and economic attention. This paper attempts to set railway and port systems serving East Africa within the broader context of ideas relating transport development with economic growth.

VIEWPOINTS ON THE RÔLE OF TRANSPORT IN LESS DEVELOPED COUNTRIES

In recent United Nations studies, transport has been described as 'the formative power of economic growth and the differentiating process' (Voigt, 1967), and attention has been directed to the fact that

transport difficulties have considerably retarded the exploitation of natural resources, industrialisation, expansion of trade . . . and in some cases the achievement of national unity.†

† *Transport development*, United Nations Economic and Social Council, New York (E/4304) 1967. See also: *Transport problems in relation to economic development in East Africa*, United Nations Economic Commission for Africa, Addis Ababa 1962.

Whilst these statements are certainly valid in general terms, as many examples illustrate, the interlinkages between the dual processes of economic growth and transport development will bear closer examination. On a world basis Berry has provided a useful general analysis of the relationship between road and rail transport network densities and the general level of economic development in a range of countries; the results suggest that some less developed countries have placed undue reliance upon selected transport media and that balanced transport development has not paralleled overall economic growth (Berry, 1960). The transport/development relationship is clearly dependent upon the specific type or range of transport media involved in a given area, upon the type of economy the transport facilities are required to serve, and upon the level of economic development at which transport media are introduced. So many variables are involved that, beyond a certain point, comparisons become relatively valueless, but theoretically there exists for a given area at a given stage of development an *optimum transport capacity* yielding efficient service without the dangers of over-capitalisation.

A distinction should be made in this context between two basic phases in the evolution of modern transport systems in less developed countries. The first phase, which may be termed *initial transport provision*, involves the construction of major rail arteries, modern roads and port facilities, and relates frequently to the earlier years of the colonial period. An outline systems is thus established which permits economic growth up to a certain point, beyond which its shortcomings begin to restrict economic development. A second phase, which may be described as *transport elaboration* and which frequently refers to the later years of colonial dependence and to a period of political independence, involves the extension of the basic system including improvements in its efficiency, which permits a higher level of economic development. The two phases are contrasted in their economic impact in the sense that the first phase obviously permits (and is likely at least initially to stimulate) economic growth, whilst the second phase is generally merely permissive in economic terms and is not the direct stimulant to development that many politicians and some economists have thought likely. Transport is the key to modern economic growth in the less developed countries in the sense that it is a *sine qua non;* but transport elaboration in an already partially developed economic system does not necessarily encourage any further growth, and indeed is unlikely to do so unless further

positive steps are taken to maximise the utility of the facilities provided by encouraging the geographical coincidence of development
projects from several economic sectors.

THE GROWTH OF RAIL NETWORKS

In the less-developed countries the decision to build a major rail
artery belongs generally to the phase of initial transport provision
and is frequently of greater long-term significance than any other
transport development. Such a railway tends to control the overall
pattern of a country's expanding transport network and thereby to
influence profoundly the pattern of its economic geography for a
considerable length of time; this is especially important in those less-
developed countries which depend largely upon a single rail artery
penetrating inland from a major port. Less significance is generally
attached to roads, branch railways and minor ports, since although
these may individually represent considerable capital investment the
basic transport system of an area is not significantly disrupted if a
decision to elaborate in a particular locality proves wrong and has to
be reversed. Only when road/rail competition over long distances
becomes intensive and a major railway loses its economic predominance, as in Ivory Coast and Sierra Leone, does the close association
between an arterial railway and the economic geography of an area
begin to disintegrate.

The geographical analysis of the changes involved in such situations constitutes an interesting area of research. Related studies have
been undertaken, for example, within the Transport Research Programme of the Brookings Institution (Fromm, 1965; Wilson, 1966).
In East Africa the impact of recent rail extensions in Uganda has
been analysed by O'Connor, 1965a. East Africa is now provided with
a railway system unified both physically and administratively, linking
the three countries of Uganda, Kenya and Tanzania. Administrative
unification was provided by the East African Railways and Harbours
Administration in 1948; all sections of the system were physically
linked by 1963. The system is based upon two rail arteries, respectively linking Mombasa with the Lake Victoria basin and Uganda,
and Dar es Salaam with Lakes Victoria and Tanganyika (Fig. 3.1).
The elaboration of the system has been outlined elsewhere (Hoyle,
1963), and is discussed here in the context of a comparative model
framework.

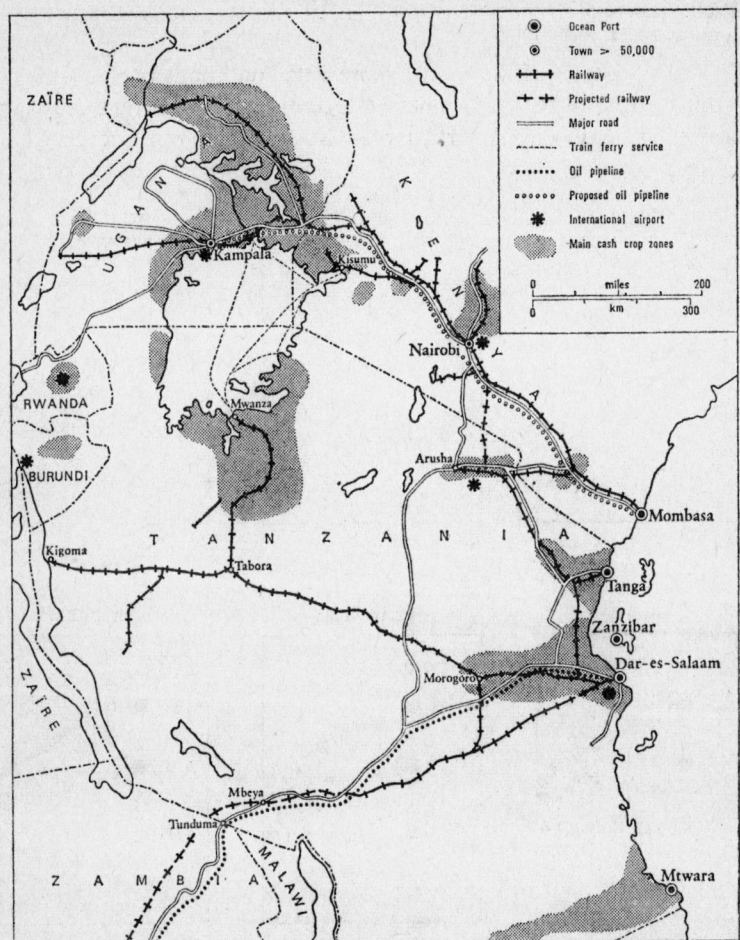

Fig. 3.1   East Africa: some aspects of transport and economic activity

A MODEL OF EAST AFRICAN TRANSPORT DEVELOPMENT

In a well-known paper, Taaffe, Morrill and Gould discussed the evolution of transport patterns in less-developed countries and proposed a model for the analysis of such patterns (Taaffe, *et al.*, 1963). Based originally upon studies in Ghana and Nigeria, the model has wide applicability. A Malaysian case study is provided by Ward,

1969. Figure 3.2 represents an application of this model to the East African case.

The first stage of the model represents conditions over a period from the first to the early nineteenth century. The diagram shows a scatter of small coastal ports, largely isolated from one another and

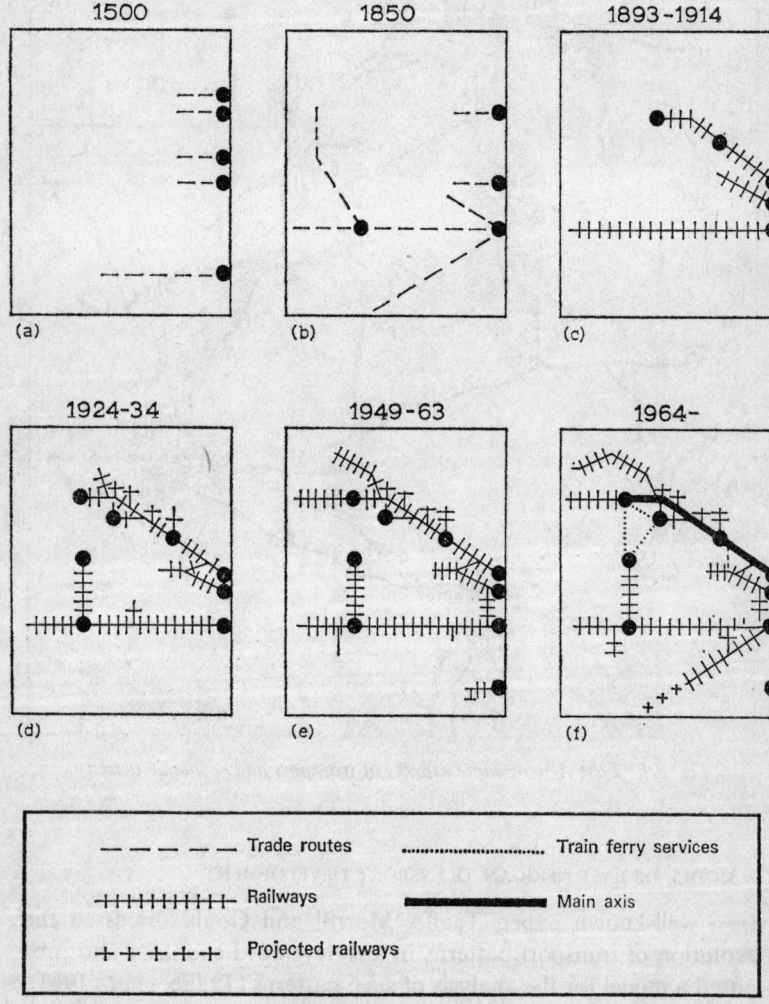

Fig. 3.2 East Africa: an application of the Taaffe–Morrill–Gould model of transport network development

linked more firmly with the maritime trading circulations of the Indian Ocean than with those parts of the interior from which slaves, ivory and other goods were regularly extracted. Although few ports or routes significant in medieval times continued to grow in later periods, the transport pattern shown in stage (*a*) underlines the historical precedents of modern systems. Stage (*b*) represents the zenith of the period of Arab trading activity based upon the island emporium of Zanzibar, from the outports of which trade routes radiated inland as far as western Uganda and Tanzania and indeed well beyond the confines of present-day East Africa. The concentration of transport routes upon Zanzibar at this time reflected the predominant role of the port in the economic and political life of the entire area, and involved a sharp decline in the fortunes and number of mainland trade centres. A more detailed discussion of pre-twentieth-century transport is contained in Hoyle, 1967*a*. The central Tanzanian transport axis initiated by the Zanzibar–Tabora slave route is re-emphasised in stage (*c*) by the Dar es Salaam–Kigoma railway, paralleled in Kenya by the Mombasa–Kisumu line. These two rail arteries, together with the less successful Tanga–Moshi line, have subsequently influenced profoundly the pattern of transport development and economic expansion in East Africa, particularly in the spheres of mineral exploitation, cash crop production, and rural and urban settlement. Since all three lines were initially based upon political rather than economic motivations, it is interesting to speculate upon the likely present-day patterns of East African economic geography, had the railways been differently located.

Stages (*a*) to (*c*) of the model thus represent the broad phase identified earlier as *initial transport provision*, with particular emphasis on stage (*c*). Stages (*d*) to (*f*) indicate successive eras in the broad phase identified as *transport elaboration*, and the routes represented are mainly rail transport routes since roads are often either parallel or locally tributary to railways, and since railways in East Africa still carry a majority of long-distance bulk consignments. Stage (*d*) saw the elaboration of the three original railways by branch- and feeder-lines designed to serve areas of increasing settlement and cash crop production, and by important extensions to Arusha, Mwanza and Kampala. Stage (*e*) is marked by the brief appearance of a railway serving the hinterland of the then newly-established deep-water port of Mtwara, by the interlinkage of the Tanga and central Tanzanian lines in the coastal zone, and by the two major extensions of the rail networks to the west and north of Uganda respectively.

These developments had varying economic relevance. The Mtwara line, designed to evacuate anticipated groundnut harvests from southern Tanzania, failed in this objective (and was lifted in 1963) partly because the new port merely supplemented rather than replaced the lighterage port of Lindi to the north, with its road-feeder system. This point is elaborated in O'Connor, 1965*b*. The principal advantage of the Mnyusi–Ruvu link is that rolling stock may be moved as required between the Kenya–Uganda and central Tanzanian sections of the system, and substantial economies result from this in view of the complementary seasonal pattern of much of the cash crop freight which is handled. The Uganda extensions, built respectively to transport copper from the west and cotton from the north, do not yet appear to have attracted a significant volume of other traffic, and together provide the main body of contemporary evidence from East Africa that transport elaboration within an established economic system is not likely to accelerate development unless further steps are taken to utilise more fully the facilities provided (O'Connor, 1965*a*). The final stage of the model indicates the intensification of surface transport along the Kampala–Nairobi–Mombasa axis (now obviously well-established as 'main street, East Africa'),† reflects the introduction of train-ferry services on Lake Victoria, and shows the contemporary transport extensions towards Zambia from Dar es Salaam. A Tanzania–Zambia railway authority was constituted in October 1968 (Griffiths, 1968). This southern branch of the East African transport network is rapidly becoming a major international axis in its own right, and is already identified by an oil pipeline and a major road; the decision to construct a Tan-zam railway based on Chinese surveys and loans was taken in November 1969. These developments are clearly related not only to the Zambian situation but also to specific agricultural development schemes‡ and potential coal and iron ore developments in southern Tanzania. Proposals for the construction of a 600 MW hydro-electric power station in southern Tanzania were also announced in 1969.

Thus a close relationship between the broad pattern of East

---

† The construction of an oil pipeline from Mombasa *via* Nairobi to Kampala is under consideration in the context of East African inter-state discussions on transport problems.

‡ On agriculture in this area, see Jatzold, 1967.

Africa's transport system and many other aspects of the geography and economy of the area. In the pre-railway stages East African trade was channelled into limited arteries linked with the Indian Ocean maritime transport system; other movements were mainly local in origin and destination. Nevertheless, these early stages provided established modes of access to and movement within the East African interior and this influenced the alignments of the pioneer railways. These in turn, with their later elaborations, are part of the infrastructure upon which the modern economy is based. In geographical terms, the specific location of the various nodes and links in the transport network (particularly the rail network) has very strongly influenced the location and economic growth of those areas that are heavily involved in the cash economy (Fig. 3.1). The transport of East Africa's bulk agricultural products (notably coffee, cotton, sisal and tea) and the export of minerals such as copper and soda ash are very largely dependent upon rail facilities. Whilst it is true that road feeder lines are important in cash crop producing areas, and that road/rail competition for transport to the ports is intensifying, the rail arteries have in a sense controlled the location of such areas. Areas of East Africa without reasonable access to rail transport but with considerable economic potential (for example parts of southern Tanzania) have been relatively slow to develop and have not generally emerged as economic core areas either during the colonial period or subsequently. In the context of the distinction made earlier between initial transport provision and transport elaboration it does not follow that contemporary transport improvements will encourage the emergence of new economic foci, but the historical relationship between transport media and patterns of economic growth is very clear.

THE CRITICAL RÔLE OF THE SEAPORT†

A very high proportion of the external trade of the less developed countries passes through their seaports (in Africa only 6 per cent is overland trade)‡ so that a seaport is particularly well-placed to act either as a growth pole or as a restrictive influence upon economic

† The general discussion in the first part of this section is based upon a more extended treatment by Hoyle and Hilling, 1970.

‡ *A survey of economic conditions in Africa*, United Nations Economic Commission for Africa, Addis Ababa (E/CN.14/397) 1967.

development. The provision of port facilities can thus be regarded, throughout the underdeveloped world, as an essential pre-condition for modern economic growth; in a less-developed country ports assume a critical role in the development process, and the stage of economic development attained is to a considerable degree a measure of the capacity and degree of sophistication of the port facilities available. The less-developed countries are today attempting to establish or renew their port systems in the context of rapid changes in the technology of maritime transport exemplified by the increasing use of containers and the increasing size of bulk carriers; thus they have the opportunity to incorporate these changes at an early stage in the process of port growth, and to ignore this opportunity would merely serve to widen the gap between the richer and the poorer countries.

East African seaports

These problems may be illustrated by the East African seaport group, which comprises five terminals handling ocean-going vessels together with a number of minor coastal ports (Hoyle, 1967*b*). Mombasa is predominant; in 1968 it handled 68 per cent of the total traffic of the mainland seaports (East African Railways and Harbour Administration, 1969). Dar es Salaam is the chief port for Tanzania, and the ports of Tanga, Zanzibar and Mtwara are relatively minor. Together, the five ports handled 14·2 million tons of shipping and 8·1 million tons of cargo in 1968.† The port group forms a dynamic hierarchy, which has experienced successive eras of concentration and diffusion of activity along the coast, and now shows a very marked emphasis upon the port of Mombasa as a centre of *sustained port dominance*. A useful discussion of Nigerian seaports in this context is Ogundana, 1970. Although the modern ports vary widely in terms of equipment and the volume of cargo handled, there is a fundamental similarity in their pattern of development which may be expressed by means of the *Anyport* model. 'Anyport' is a hypothetical port which represents the common experience of the port group as a whole (Hoyle, 1968). From its original function as a shelter for Arab sailing dhows trading around the Indian Ocean, *Anyport* in East

† These figures represent a record level of traffic flow through the port group as a whole, in spite of the problems arising from the continued closure of the Suez Canal. (See East African Railways and Harbour Administration, 1969).

Africa has successively developed primitive jetties which were re-placed by lighterage quays, superseded in turn by deep-water quays to which are now added the specialised facilities for bulk oil and grain cargoes.

Problems of congestion and cargo-handling techniques

Two main problems affect the further development of the East African seaport group, and both are closely related to the techno-logical, political and economic factors which have influenced the emergence of the port group in its physical and historical setting. The first, which has been discussed elsewhere, is that of equating available facilities with actual and estimated traffic demands, together with a range of related questions involving congestion in the ports, the seasonal nature of traffic flows, the application of capital resources and the effects of delays upon economic development (Hoyle, 1967b). The basic question is that of acquiring and applying capital re-sources at the right time and in the most profitable manner. At both Mombasa and Dar es Salaam extensions to the available deep-water quayage have recently been made, stimulated at Dar es Salaam by increasing Zambian traffic.† The second is the problem of the adaptation of East African ports to modern techniques of cargo handling, including the increasing unitisation of cargo and the use of containers, the elaboration of specialised cargo-handling facilities, and the physical amelioration of the harbours in order to allow the entry of large bulk-carriers.

The less-developed countries have been faced in recent years with the need to consider to what extent it is in their interests to participate in the 'container revolution' in transport technology. World trends towards the increasing unitisation of cargo movements, stimulated by the need to increase efficiency and reduce costs, have not left ports in the less-developed countries unaffected since all ports serve the same world fleet of shipping and must provide at least a minimum range of similar facilities in order to attract traffic. Today the pattern of inland transport in East Africa is characterised by minimal unitisation, and although pallets have been extensively used in the ports for twenty years there is little sign as yet of the more widespread use of con-tainers. Part of the reason for this situation is the relatively small

† Four additional deep-water berths (nos. 16–19) have been built at Mombasa, and at Dar es Salaam three new deep-water berths (nos. 4–6) have been brought into service.

volume and the very varied nature of East African trade, dominated by bulk agricultural exports and imports of consumer goods; anxieties have also been expressed about the reduced labour requirements which unitised systems would involve in the context of East Africa's chronic unemployment problems. A large proportion of East African exports are already efficiently handled in bulk and relatively few benefits would accrue from further unitisation; but containerised import cargoes receive a 20 per cent rebate on port charges, and National Trading Corporations may encourage the unitisation of imports by establishing wholesale distribution centres inland. The first containerised cargoes despatched from Britain to an inland East African destination arrived in July 1968 and were handled by East African Containers Ltd. (a subsidiary of the Express Transport Company).

Existing deep-water berths in East Africa are suitable for the handling of some forms of unitised cargo although problems are likely to arise from the shortage of transit space near the quays. It may be argued that the use of containers might effectively reduce congestion at Mombasa and Dar es Salaam without necessarily involving any further immediate increases in berth accommodation, and in this respect whilst new methods complicate the forward planning of port development they nevertheless improve considerably the long-term prospects of the deep-water terminals. Prospects for the three smaller East African terminals are less assured, however, since, although Mtwara may secure a share in Zambian traffic† and benefit from the growing emphasis on southern Tanzanian development, the possibilities for expansion at the lighterage ports of Tanga and Zanzibar are limited and their traffic seems likely to suffer a continued relative decline as the very uneven distribution of East African maritime trade between the five seaports becomes increasingly marked. In 1968 Mombasa and Dar es Salaam together handled 94·8 per cent of the total cargo traffic of the four mainland seaports (East African Railways and Harbours, 1969).

With these considerations in mind, it would seem to be in East Africa's interests to develop an intermediate technology rather than to attempt to establish a complete container programme which the area cannot afford and which is in any case inappropriate in the present stage of development. The basis of successful container

† Zambian copper exports reached Mtwara by road (*via* Tunduma) from September 1967, but ceased in late 1968 due to poor road conditions.

operations in advanced economies is the relatively high-value two-way flow of manufactured goods in quantity. This is not yet relevant to the East African traffic situation, but palletisation can improve efficiency considerably without the radical changes and heavy capital expenditure which containerisation would involve. The application of methods of unitisation must be based on local circumstances, and in East Africa these indicate that a modest capital outlay on cargo palletisation and pre-slinging equipment may be worthwhile but that the construction of specialised container berths (at an estimated cost of £2 million to £3·5 million each) is not yet an economic proposition. Nevertheless, the port authorities must clearly watch their investment programme with great care so that facilities developed for use by conventional ships can be converted for use by specialised container vessels at a later date with minimum disruption and expense. Three of the deep-water berths recently built at Mombasa (see footnote on page 59) are designed to facilitate adaptation as container berths.

CONCLUSION

Hance (1967*a*) has emphasised 'the great power of improved transport to quicken the economic pulse of a region', and it is clear that all transport media are vital factors in the economic growth of the less developed countries. In most such countries transport systems have developed without any overall plan, and frequently the construction of a road or a railway has been a political issue rather than an economic matter. Today the shortcomings of transport systems so haphazardly developed are everywhere apparent, but the developmental role of transport is more complex and more sensitive than has commonly been stated. Transport media are permissive rather than stimulative in themselves, but potentially they may effect widespread economic transformation in association with other development schemes. In the case of East Africa, as in many other less developed areas, 'the goal should be integrated development' (Hance, 1967*b*). If the railways and ports are to continue to facilitate and not to discourage economic expansion, a continuing need exists for heavy expenditure on the improvement and adaptation of existing facilities, and for the integration of transport planning within the East African Economic and Social Community. In 1969 the governments of Kenya, Tanzania and Uganda commissioned an East African

Transport Study to provide 'recommendations . . . to co-ordinate the use and development of the various modes of surface transport . . .', and also secured an £11·4 million loan from the World Bank for improvements in bulk cargo handling facilities at Mombasa and Dar es Salaam.

REFERENCES

BERRY, B. J. L. (1960). An inductive approach to the regionalization of economic development, *Research Paper 62*, Department of Geography, University of Chicago.

EAST AFRICAN RAILWAYS AND HARBOURS ADMINISTRATION (1969). *Annual report for the year ending 31st December 1968*, Nairobi, p. 66.

FROMM, G. (ed.) (1965). *Transport Investment and Economic Development*, Washington.

GRIFFITHS, I. L. (1968). Zambia's links with East Africa, *E. Afr. Geog. Rev.* 6, 87–9.

HANCE, W. A. (1967a). *African Economic Development*, New York, p. 117.

HANCE, W. A. (1967b). *Ibid.*, p. 118.

HOYLE, B. S. (1963). Recent changes in the pattern of East African railways, *Tijd. Econ. Soc. Geog.* 54, 237–42.

HOYLE, B. S. (1967a). Early port development in East Africa: an illustration of the concept of changing port hierarchies, *Tijd. Econ. Soc. Geog.* 58, 94–102.

HOYLE, B. S. (1967b). *The Seaports of East Africa*, Nairobi.

HOYLE, B. S. (1968). East African seaports: an application of the concept of Anyport, *Trans. Inst. Brit. Geog.* 44, 163–83.

HOYLE, B. S. and HILLING, D. (eds.) (1970). *Seaports and Development in Tropical Africa*, London.

JATZOLD, R. (1967). *The Kilombero Valley*, Weltforum Verlag, for Ifo Institute for Economic Research, Munich.

O'CONNOR, A. M. (1965a). *Railways and Development in Uganda*, Nairobi.

O'CONNOR, A. M. (1965b). New railway construction and the pattern of economic development in East Africa, *Trans. Inst. Brit. Geog.* 36, 21–30.

OGUNDANA, B. (1970). Patterns and problems of seaport evolution in Nigeria, (eds. Hoyle and Hilling), *Seaports and Development in Tropical Africa*, London, pp. 167–82.

TAAFFE, E. J. *et al.* (1963). Transport expansion in under-developed countries: a comparative analysis, *Geog. Rev.* 53, 503–29.

VOIGT, F. (1967). *The Importance of the Transport System for Economic Development Processes*, United Nations Economic Commission for Africa, Addis Ababa (E/CN. 14/CAP/3G).

WARD, M. J. (1969). Progress in transport geography, (eds. R. U. Cooke and J. H. Johnson), *Trends in Geography*, London.

WILSON, G. W. (ed.) (1966). *The Impact of Highway Investment on Development*, Washington.

4 The Search for Spatial Regularities in the Development of Australian Seaports 1861–1961/2

PETER J. RIMMER

GEOGRAPHICAL studies of seaports have been concerned in the past with such varied topics as physical layout, the origin and destination of cargo and character of shipping handled (see, for example, Weigend, 1956; Boxer, 1961; Bird, 1963). With their emphasis on the uniqueness of individual ports or the peculiar features of groups of centres, these studies have paid little attention to the problem of providing a model against which the various aspects of port geography can be measured. The first notable departure from the traditional method of studying seaports was made by Bird who evolved the concept of 'Anyport' for comparing developments in the layout of major British ports since their inception (Bird, 1963). Then Taaffe, Morrill and Gould in their study of the expansion of transport networks in underdeveloped countries highlighted first, the evolution of the spatial patterns of port locations with the improvement in internal accessibility and secondly, the accompanying process of dominance ranking and the emergence of a port hierarchy (Taaffe *et al.*, 1963). However, in their emphasis on the development of landward connections Taaffe *et al.* neglected changes in the organisation of maritime space, which is as essential to the development of a port as land transport.

In an attempt to incorporate both the changes in the maritime and landward transportation networks a simple model was developed by the author, based on the assumption that a general process of dominance ranking has taken place through the improvement of internal accessibility (Rimmer, 1967a). The model is intended to serve as a yardstick for comparing changes in the evolution of seaports. Its validity was tested initially by applying it to the distribution of New Zealand seaports at six different dates between 1853 and 1960. In this paper a refined version of the first model is put forward (Fig. 4.1). The refined model incorporates the weighting of transportation networks to provide a clearer appreciation of the process of dominance ranking and also introduces the aberrant case of a port that survives the hierarchical process to provide a closer simulation of reality. Its utility outside New Zealand is tested by applying it to the distribution of Australian ports at six different dates between 1861, when the first detailed figures for the whole of Australia were available, and 1961–2.

DEVELOPMENT OF AUSTRALIAN SEAPORTS

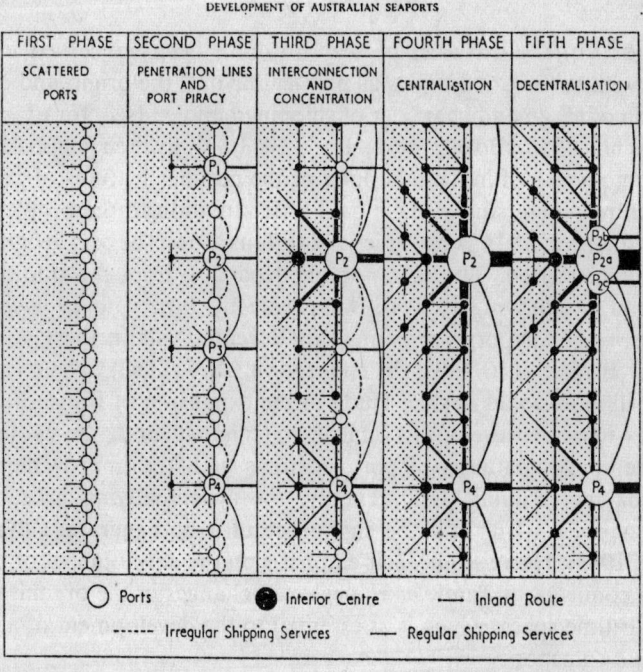

Fig. 4.1   The idealised-type sequence of Australian port development

Unfortunately, as in New Zealand, there is a marked paucity in Australia of suitable measurements for gauging a port's status over such a lengthy period of time. Indeed, the only measurement available for the study period to assess the changing status of Australian ports is the net registered tonnage of shipping entering individual ports. The use of this standard over 100 years is not altogether satisfactory as the Plimsoll Line—which regulates the weight of cargo a ship can carry—was raised some years ago to allow greater cargo to be carried without altering the net registered tonnage figure (Kendall, 1948; Neale, 1955; Rimmer, 1966). In spite of this drawback, net registered tonnage provides a useful guide—in the absence of alternative yardsticks—for distinguishing the movements of seaports within the idealised-type sequence of port development. The conclusions made on such a basis, however, can only be suggestive and cannot in any way be definitive.

THE SEQUENCE OF PORT DEVELOPMENT

Five phases are recognised in the idealised-type sequence of port development (Fig. 4.1). The first phase consists of a dispersed pattern of seaports with centres scattered along the coast serving limited hinterlands. Irregular visits of trading vessels provide the only connections between ports. With the emergence of the main lines of penetration into the interior during the second phase of port development certain ports are able to expand at the expense of others, such as P_1, P_2, P_3 and P_4. Port concentration is accentuated because P_1, P_2, P_3 and P_4 develop as the foci for four separate networks of routes. The feeders continue to develop until P_1, P_2 and P_3 are linked together. As it occupies a central position P_2 expands its influence by pirating the trade of P_1 and P_3 in the third phase of the idealised sequence. The development of ports P_1 and P_3 slackens and they revert to their former status. The port P_4, however, survives because it is not linked to the system. When the link up between the two separate systems is completed in the fourth phase of the idealised-type sequence the port P_4 has gained sufficient momentum not to be overshadowed completely by port P_2, despite the centralisation of economic activities on the latter port. Port P_4 is able to withstand the competition by providing limited or specialised services for its immediate hinterland. Further growth, however, is confined largely to port P_2. Indeed, the continuing expansion of the transport network and the intensification of economic activities within its immediate hinterland overtaxes its capacity and the fifth phase is characterised by decentralisation at the major port. Port P_2b and P_2c are established to provide specialist functions enabling the initial port P_2 (now P_2a) to concentrate on general cargo services.

It is probably more realistic to think of the entire sequence of port development as a process rather than a series of distinct temporal phases. Thus, at any given point in time the pattern of Australian seaports may reveal the co-existence of all four phases. It is convenient, however, to consider the sequence in phases.

The first phase: scattered ports

The first 75 years of European settlement in Australia produced a rash of scattered ports serving limited hinterlands. Consequently, when the first port statistics were published for all Australian States in 1861 there were already 29 overseas ports serving a population of 1·2 m (Fig. 4.2). This dispersed pattern of ports concentrated at particular localities around Australia's 12,300-mile coastline reflects the genesis and subsequent development of the continent.

The first settlement was established at Sydney Cove, Port Jackson, in 1788 as a penal colony for United Kingdom convicts. Its site was chosen because 'ships can anchor close to the shore so that at very small expense quays can be made at which the largest ships can unload'.† This emphasis on the possibilities of providing port facilities was critical to a fledgling colony as ships provided the only contact with the mother country over 11,000 miles away. Of necessity, port facilities also figured prominently in the decisions to develop new settlements separate from Sydney because the only method of inland penetration was by bullock wagons which were both slow and costly even for short distances. As water access to the interior of eastern Australia was inhibited by the absence of long navigable rivers or the possibilities of building canals, movements of goods over 30 miles were confined invariably to the salt-water highway. (Much of this section has been inspired by Blainey, 1966a.) Thus, the establishment of additional settlements in Australia either to provide penitentiaries for convicts committing offences within the Colony (Newcastle 1803, Port Macquarie and Brisbane 1824), or to safeguard other parts of Australia's vast coastline from possible French settlement—Hobart (1803, took convicts 1804), Launceston (1806), Fort Dundas (1824), and Raffles Bay (1827), both near Darwin, Westernport Bay (1826), King George Sound (1827)—or alternatively to provide shore bases for whalers (Eden, Port Fairy, Portland, Port Eliot, Esperance and Macquarie Harbour) had to proceed by sea.

The further development of settlement and the exploitation of Australia's natural resources by free colonists—largely independent

† 'Governor Phillip to Lord Sydney,' *Historical Records of Australia*, ser. I, Vol. I, pp. 17–19. Quoted by J. Bird, 'The Foundations of Australian Seaport Capitals,' *Economic Geography*, **41**, 1965, 283–99.

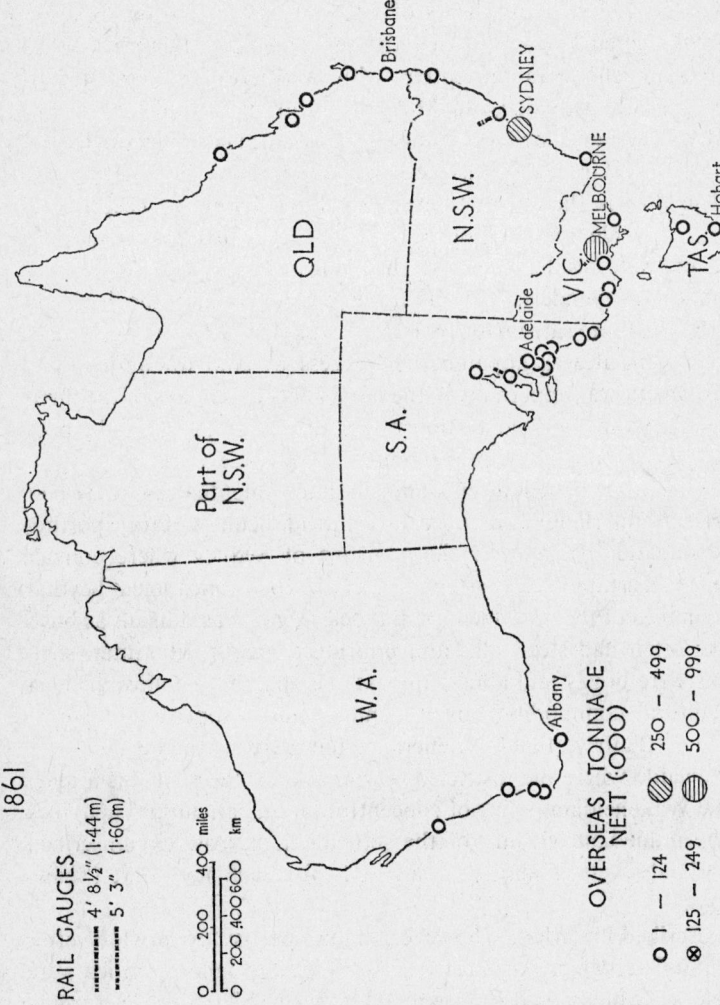

Fig. 4.2 Australian seaports, 1861

of Sydney's control—were also geared to sea transport. Access to port facilities near the entrance of rivers was a prime factor determining the location of the attempts at free colonisation in Perth (1829), Melbourne (1835) and Adelaide (1836). Indeed, the survival of the colonies during the early years of settlement depended on sporadic visits of United Kingdom ships and the irregular services provided by top-sail schooners and cutters with other colonies in Australia. Neither Perth, Adelaide, nor Melbourne, however, could match the depth of water and shelter provided in the natural harbours of Hobart and Sydney. Lightering was necessary from ships anchored in the roadstead or in the centre of the river almost from the inception of Perth, Adelaide and Melbourne as ports. The subsequent development of trade resulted in costly harbour works and dredging programmes being undertaken so that the ports could meet the demands of shipping more satisfactorily. Even at this early stage in Australia's development it was apparent that access to land for agricultural development was uppermost in the minds of the founders of the three settlements for free colonisation rather than the quality of the port which was capable of improvement by man-made works.

Irrespective of their origins and ultimate potential, all ports had restricted hinterlands at first because of the difficulties of transporting goods inland. Except for the environs of Sydney where convict labour had established a system of turnpike roads, movement beyond the confines of the settlement by bullock wagon was difficult as bush tracks often had steep hills and unbridged creeks. In winter, such tracks were boggy and almost impassable, making the movement of agricultural commodities an even more tedious operation than in summer (Blainey, 1966*b*). Generally, therefore, with the exception of Adelaide where the restricted service area of the port was a tenet of the Wakefieldian policy of concentration on agricultural activities in the immediate vicinity of the settlement, narrow circumscribed hinterlands were a matter of necessity for each new port (Shaw, 1960*a*).

Proscribed hinterlands, however, did not persist for very long after the initial settlement of Australia because economic activities were developed which could overcome the cost-disabilities of dray transport and survive in the interior beyond the river flats and plains near the coast.

First, there was the development of sheep farming, geared to producing wool for export to Europe. The price of wool was so high that, unlike an agricultural commodity such as wheat, it could survive not only the cost of a trip overland by bullock wagon from the interior to the port, but also the price of transporting the wool from an Australian port to Europe for consumption in the expanding, mechanised woollen industry. Originating in the area around Sydney pastoralism extended via the convict road to Bathurst (completed 1815) over the Blue Mountains to the inland plains beyond. So successful was the policy of squatting on the interior grasslands that sheep farming was extended from the mainland to Tasmania in the mid-1830s and from there back to the mainland through Melbourne (1835). The surge of pastoralists emanating from the Port Phillip settlement at Melbourne met their counterparts moving south from the Sydney settlement. By 1861 sheep farming 'occupied' in a rather sporadic fashion a huge unfenced area of land extending from the back of Brisbane (Queensland) inland to Port Augusta (South Australia) (Shaw, 1960*b*; Perry, 1963; Blainey, 1966*c*).

Grafted onto the sheep economy from 1851 onwards was goldmining. A wave of discoveries in New South Wales and Victoria triggered-off a series of rushes in the early 1850s. Victoria soon outstripped New South Wales and Ballarat and Bendigo became two of the main centres of the industry. Gold, like wool, could withstand the high cost of being moved by dray from inland locations. As roads to the goldfields were improved, an increasing number of horses were introduced to complement the slow-moving bullocks. Further, from 1854 onwards, in Victoria, horses were also used by Cobb and Company to provide fast passenger transport by coach. Such was the development of goldmining that it quickly overcame transport disabilities and throughout the 1850s and 1860s it was Australia's main export. An excellent description of the goldmining era in Victoria is given by Serle, 1963.

By 1861 settlement, geared to producing gold and wool for export, was firmly established. Responsible government had been introduced in New South Wales, Victoria (1855), Tasmania (Colony 1825, responsible government 1856), South Australia (1856) and Queensland (1859). Even Western Australia (Colony 1829) after a dismal start had made considerable progress and the last convicts were shipped to Perth in 1860. The whole prosperity of the newly formed colonies focused on the ports which functioned to export wool and gold overseas and to import supplies and raw materials for the growing industries and the inhabitants of the coastal cities and the goldminers, squatters and bushworkers in the interior. Thus, the pattern of Australian ports in 1861 reflects the imposition of the development of gold and wool economy on to the surviving remnants of the early period of 'foothold settlement' geared to providing gaols, military outposts and havens for whalers.

The coastal necklace of ports in 1861, separated from each other by distances ranging from 100 to 1,500 miles, is graphically described by Blainey as having much lace and a few beads haphazardly arranged (Blainey, 1966*d*). Gaps in the necklace were filled up in succeeding years as development beyond the settled years in a country with only 242 miles of railway inevitably had to proceed via the coast (Table I). In 1871 the number of overseas ports had increased by six, compared with 1861, to 35 (Fig. 4.3). There were also 11 coastal ports transshipping goods to and from the main ports as well as five river ports on the Murray, which was Australia's only inland water system capable of navigation far from the coast.

Table I. Australia: Railway
Route Mileage Opened
(Public and Private)

1861	243
1871	1,042
1881	4,192
1891	10,123
1901	13,551
1911	18,012
1921	26,202
1931	27,668
1941	27,956
1951	27,602

Source: Department of National Development, *Atlas of Australian Resources: Railways*, Canberra, 1954.

This increased scatter of ports reflected the search for gold and the continuing advance and development of pastoralism. The search for and discovery of gold resulted in new ports being established in Queensland and the expansion of sheep farming stimulated an increased use of the Murray's 2,000-mile river system by paddle steamers. The first paddle steamer went up the Murray River from South Australia in 1853 (Blainey, 1966*e*). Despite its treacherous mouth,

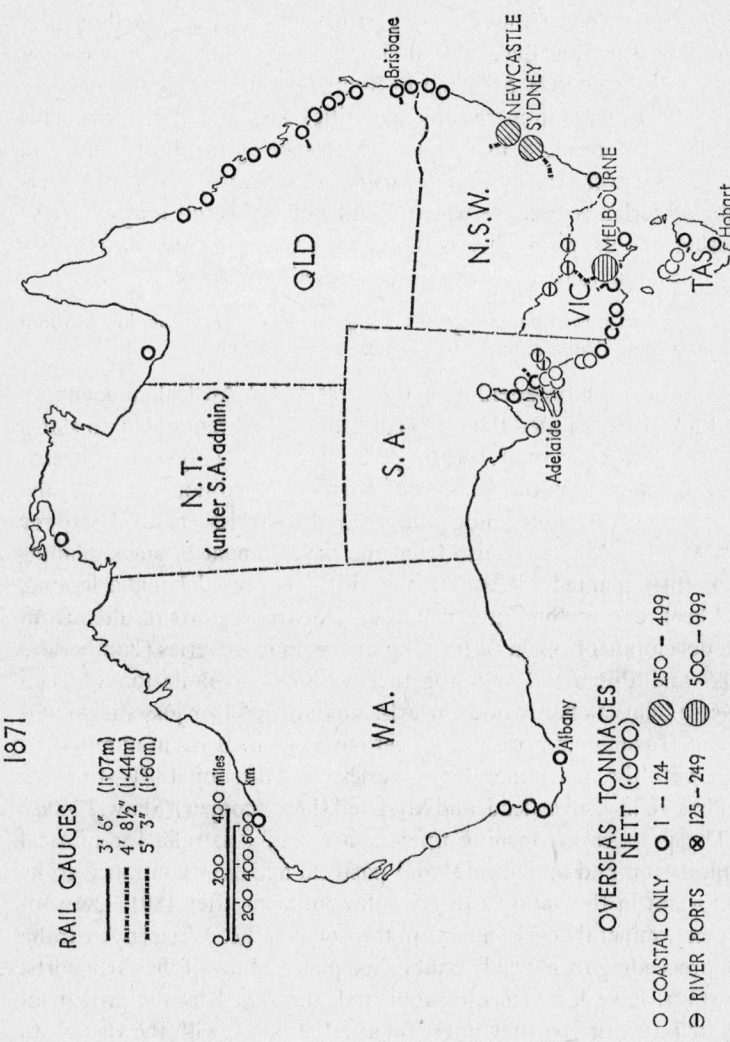

1871

RAIL GAUGES

3' 6" (1·07m)
4' 8½" (1·44m)
5' 3" (1·60m)

200   400 miles
km
200 400 600
200 400 600
0

OVERSEAS TONNAGES
NETT ('000)

○ COASTAL ONLY   ○ 1 – 124   ◐ 250 – 499
⊖ RIVER PORTS   ⊗ 125 – 249   ◉ 500 – 999

W.A.

N.T.
(under S.A. admin.)

S.A.

QLD

N.S.W.

Albany

Adelaide

VIC

MELBOURNE

NEWCASTLE
SYDNEY

Brisbane

TAS

Hobart

Fig. 4.3 Australian seaports, 1871

snags and shallow draught, the River Murray was used to move as much as one-third of New South Wales wool clip in the 1860s. Steamers carried the wool either to Adelaide or the railhead at Echuca (established 1864) for export overseas through Melbourne.

The outstanding development in the establishment of ports between 1861 and 1871 occurred in South Australia where wheat was exported overseas in increasing quantities from 1866 onwards. This development resulted in a glut of ports being established in and around Spencer Gulf in South Australia each with their own watertight hinterlands because wheat could not withstand long journeys overland by dray.† Such was the pace of the establishment of ports that the *Jamestown Review*‡ was prompted to remark.

'At the present rate of building jetties a map of our gulf coast would show in a few years more than a remote resemblance to a . . . comb. . . .'

Elsewhere the broadening of the basis of the Australian economy after 1871 necessitated the establishment of further ports in all states until there were as many as 70 operating in 1891 (Fig. 4.4). Queensland's expansion in the number of ports stemmed not only from the extension of sheep-farming and gold discoveries (main discovery Mt. Morgan 1882) but also from the development of sugar plantations (first planted 1862) with the aid of imported kanaka labour, and beef-rearing. Similarly in Western Australia, ports resulted from the development of sheep farming and gold discoveries (Kimberleys 1885 and Pilbarra 1888), together with the exploitation of such diverse items as hardwoods, pearls, sandalwood (for joss sticks) and guano. Tasmanian ports also experienced an increase in numbers to cope with the new mineral discoveries (Mt. Bischoff 1871—tin, Mt. Zeehan 1885—silver/lead, and Mt. Lyell 1886—copper) (Shaw, 1960*a*).

The practice of opening up new areas of Australia for mineral exploitation and agricultural and pastoral activities continued to be important in the nation's frontier development after 1891. Even today new mineral developments in the northwest of Western Australia are proceeding from newly established ports. Many of the early ports, however, have long since disappeared, destroyed by the growth of major ports, or else they linger on as relict ports, with the visit of an occasional vessel carrying bulky commodities to remind them of

† D. W. Meinig (1963), *On the Margins of the Good Earth*, London; especially pp. 124–65.

‡ *The Jamestown Review*, Aug. 7, 1879. Quoted by Meinig, p. 124.

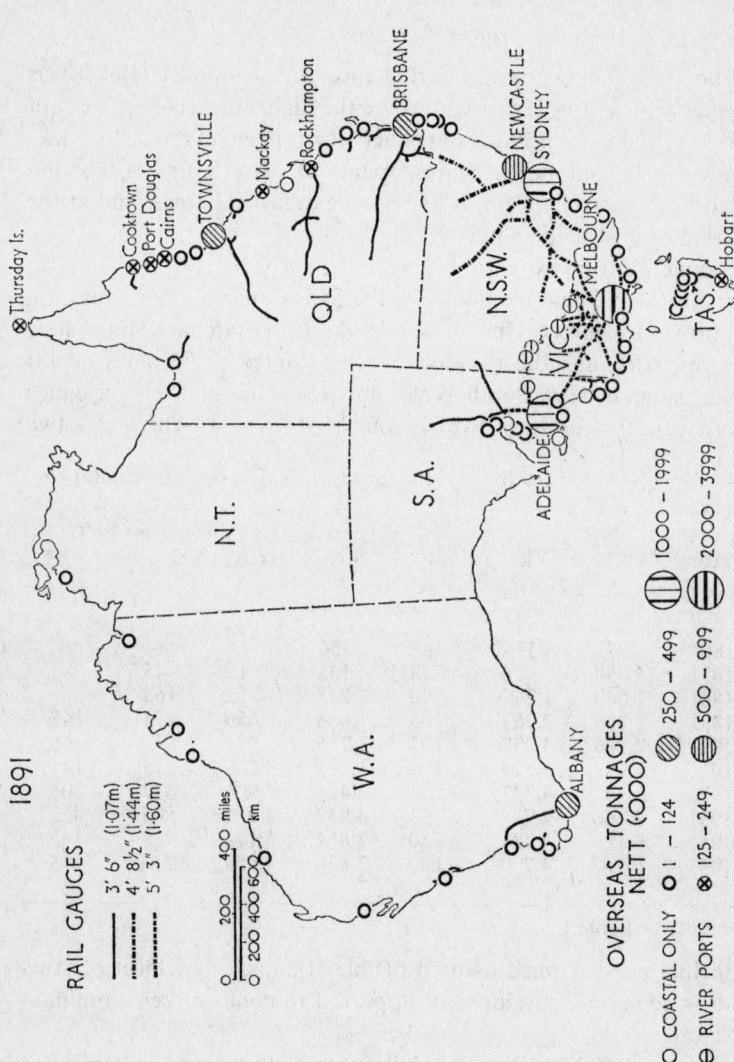

Fig. 4.4 Australian seaports, 1891

their former importance as the main inlet and outlet for a limited hinterland.

The second phase: penetration lines and port piracy

All ports had in theory an equal chance of development. But it was already apparent in 1861 that despite the highly dispersed pattern of ports there was a concentration of activities on two of them (Fig. 4.2). Sydney and Melbourne, State capitals of New South Wales and Victoria respectively, centres of growing industrial areas, and at the starting-points of the routes reaching areas of mineral exploitation and rapid pastoral expansion inland, had emerged as the two leading ports in 1861. They functioned as the foci of maritime organisation by providing direct United Kingdom, inter-state and intra-state services. Although the railways only amounted to 73 miles of 4 ft 8½ in. gauge in New South Wales and 114 miles of 5 ft 3 in. gauge in Victoria the future of Sydney and Melbourne as Australia's two

Table II. Railways Route Mileage Opened (Public and Private)

Year	N.S.W.	Vic.	Qld.	S.A.	W.A.	Tas.	N.T. and A.C.T. only
1861	73	114	—	56	—	—	—
1871	358	276	218	133	12	45	—
1881	1,040	1,247	800	845	92	168	—
1891	2,263	2,763	2,205	1,666	656	425	145
1901	2,926	3,238	2,904	1,736	1,984	618	145
1911	4,027	3,574	4,390	1,993	3,208	675	145
1921	5,402	4,337	7,013	3,463	4,906	877	204
1931	6,160	4,742	6,795	3,932	4,911	806	322
1941	6,196	4,784	6,750	3,861	5,112	758	495
1951	6,167	4,700	6,695	3,856	4,959	730	495

Source: As Table I

principal ports seemed assured (Table II). Each new facet of Australia's economic development appeared to confirm their dominant position.

The dominance exerted by Melbourne and Sydney was confirmed in 1871 by the penetration of railway lines into the interior (Fig. 4.3, Table II). Sydney was linked to Goulburn (1869) 131 miles away and Melbourne had rail access to the Ballarat (1862) and Bendigo (1862) goldfields and to Echuca, a port on the Murray River. The latter connection enabled Melbourne to capture a fair proportion of the wool trade originating in the Riverina which had hitherto been carried down to the Murray for export through South Australia (Smith, 1962, 1964). Consequently, the port was able to move ahead of Sydney and become Australia's leading port in 1871 (Fig. 4.3).

The hierarchical pattern of ports, however, was more complicated in 1871 as Newcastle had emerged as the third major port. Serving an inland pastoral area distinct from Sydney, Newcastle was also linked by rail to the Maitland Coalfield. It was this latter activity that was responsible for the port's growing importance because it served as a coaling station for overseas vessels and an increasing number of coastal steamers.

This incipient concentration of activity on a limited number of ports recognised in 1861 and 1871 was made even more apparent in 1891 by the impetus derived from the development of railway penetration lines. Between 1871 and 1891 the length of railways increased from 1,042 miles, to 10,123 miles (Table I). The railways were not in a connected network. Most ports developed their own railway connections with their hinterlands long before interconnecting lines were contemplated, much less completed. The general layout was of short lines running from a port to an island settlement separated from the next system by as little as half a mile and as much as several hundred miles. Indicative of the character of development were the six unconnected railway lines in Queensland running from a port to inland destinations. The Queensland ports with these links were Rockhampton (1867), Townsville (1880), Maryborough (1881), Bundaberg (1881), Mackay (1885), Cairns (1887), Cooktown (1887), Normanton (1889) and Bowen (1890).† All of these lines were on 3 ft 6 in. gauge, which had been introduced in Queensland (1865), Tasmania (1871) and Western Australia (1873) because smaller gauge railways were cheaper to construct than larger ones (Table II). South Australia also adopted the 3 ft 6 in. gauge in addition to the 5 ft 3 in. gauge. New South Wales persisted with the English 4 ft 8½ in. gauge and Victoria with the Irish 5 ft 3 in. gauge.

The expansion of the land transportation network reacted on the organisation of maritime space also as regular steamer services were established by Howard Smith Ltd., Australian United Steam Navigation Co. Ltd., Adelaide Steam Ship Co. Ltd., McIlwraith and

† E. C. Chapman, The Ports of Central and North Queensland, Unpublished MS prepared for ANZAAS (Section P) meeting at Brisbane, 1961 (duplicated).

McEacharn Pty. Ltd., Union Steam Ship Co. of New Zealand Ltd. to serve the ports with connections with the interior. A network of services stretched from Cairns to Fremantle and between the mainland ports to Tasmania. With the concentration of economic activity on the starting-points to inland locations the port authorities endeavoured to facilitate the loading and discharge of cargo by providing wharves, warehouses and improved depths to meet the growth in the size of ships. In Melbourne, for instance, the first pier constructed was Gem Pier, Williamstown (1839). Additional piers were added in subsequent years and depths were improved from 13 ft in 1854 to 18 ft in 1862. It was not until the establishment of the Melbourne Harbour Trust that a co-ordinated and planned port emerged. With the aid of an eminent English engineer, Sir John Coode, the river was shortened by a cut (Coode Canal) and depths improved to 23 ft. Subsequent dredging has now deepened the fairway to 36 ft.†

The emergence of improved ports and the development of penetration lines generated a series of spatial readjustments because the comparative location advantages of ports shifted by 1891 (Fig. 4.4). In addition to Melbourne and Sydney, Adelaide was now an important port. Newcastle, however, had failed to maintain its growth following the link-up with Sydney's railway network and was ranked in the second tier of ports together with Townsville, Brisbane, and the coaling station of Albany. Many of the smaller ports, in contrast, lost their function in the external trade because their positions were usurped when they were linked to the network of larger ports. Only those ports beyond the railway network retained their importance. Thus, though the railways were primarily intended as a means of opening-up the country to settlement, railway communications had in effect given the port with a line an advantage over its neighbours without one for road transport beyond the main towns by bullock, horse or camel was still relatively slow, hazardous and costly.

The importance of obtaining rail connections is highlighted in the respective development of Melbourne and its potential rivals Portland and Geelong. Rail links confirmed Melbourne's dominance as the main Victorian port. Its central position, superior harbour facilities and control over the State finances enabled it to outstrip Portland and Geelong.

Portland was settled as early as 1834 and developed as the outlet

† Based on information supplied by the Melbourne Harbour Trust, Melbourne, Victoria.

for squatters in the Western District following the first land sales in 1840. (This section on Portland is derived from Learmonth, 1960*a*.) Although the port had a 150 ft pier in 1859 its subsequent development was smothered by the State Government depriving it of necessary money to finance harbour works. The proposed venture of connecting the port to its hinterland by tramway failed. Indeed, the development of the port's rail links proceeded slowly. The link to Hamilton (53 miles) was not completed until 1877. Further connections to Mt. Gambier and the Wimmera were not effected until 1917 and 1920 respectively. Even when some connections were finally made, differential freight rates favoured Melbourne at the expense of Portland. Such practices irked local opinion and provoked the local Member of Parliament to comment:

'I take it that these differential freight rates are a form of protection to protect Melbourne from people who have the audacity to enter into competition with them for one small part of the export trade' (Learmonth, 1960*b*).

Such was the progress of Portland that, prior to the port works in 1902 it could only handle vessels of under 250 tons.

The concentration of activity on Melbourne also accounted for Geelong's slow progress. A shallow bar at the entrance to the channel into the inner harbour was not dredged until 1881. This necessitated the lightering of cargo 8 miles from the roadstead to the port. In the meantime a rail link was completed in 1856 between Geelong and Melbourne where cargo could be handled directly onto the wharves and railed to Geelong and Ballarat. As a result, imports fell by three-quarters between 1855 and 1865 (McKenzie, 1918).

Thus, where no port enjoyed the ideal combination of an easily accessible hinterland and a safe deep-water harbour, the government exerted a persuasive influence through its public works programme. This influence was not confined to Victoria, because in South Australia the government decided which harbours to improve and which railways to build. Meinig indicates the competition for government favours with another apt quotation from the *Jamestown Review*:

'One not uncommon dodge is to get the Government to erect a wharf or jetty . . . at the head of a shallow creek. They are next told that the outlay is utterly useless without the channel be cleared and deepened so that vessels may come up to load. This being done the discovery is made that this famous jetty or wharf being at last approachable from the sea is still inaccessible by land, and a railway is the only thing that will put things to rights.'†

† *Jamestown Review*, Aug. 7, 1789. Quoted by Meinig, pp. 161–2.

Railways were, therefore, the prime arbiter of a port's relative status.

Few ports with single penetration lines survive today because the subsequent process of interconnection deprived them of their water-tight hinterland. Darwin with its line to Pine Creek (1889 extended to Birdum 1918) is one of the few remaining examples. This phase of port development with a single line penetrating a port's hinterland is likely to occur again in Western Australia as a means of exploiting the mineral deposits in the north-west of the State.

The third phase: interconnection and concentration

Penetration was followed by lateral interconnection as feeder networks of different port systems joined. This process was already well-advanced in Victoria, New South Wales, and South Australia by 1891 (Fig. 4.4). It permitted Sydney, Melbourne and Adelaide respectively to attract shipping at the expense of neighbouring ports. All three port systems were linked together by rail—although there was a break of gauge between the Victoria and New South Wales systems (met at Albury 1883). Such a break was of little importance as each Colony functioned as a separate entity (Blainey, 1966*f*). The main economic connections of the Colonies were with Europe than with each other and their ports functioned largely to ship wool and wheat and, since the success of the first refrigerated vessel in 1879, increasing amounts of dairy products and fruit. Inter-colonial linkages were not strong because custom barriers prevented the free flow of commodities.

The dominance of these ports was far from complete in 1891 (Fig. 4.4). Adelaide's attempt to usurp the functions of South Australian ports was blunted by the orientation of railways at right-angles to the coast and the differences in railway gauges. Melbourne's rivals in Victoria were less well-placed as the port's radiating railways reached out and tapped the hinterlands of its smaller competitors. Sydney, however, had to contend with Newcastle, which had developed into an important port during the time it operated as the focus of its own railway network. The completion of the bridge across the Hawkesbury River (1889) deprived Newcastle of some of its direct overseas imports but the port still continued as the main outlet for northern New South Wales and its own local coal industry.

During the period of isolation Newcastle had gained sufficient impetus, therefore, to withstand the link-up with Sydney and resist possible absorption and reduced status. It is to explain the survival of a port, such as Newcastle, after the links between a smaller and larger port have been forged that the aberrant case was introduced in the idealised-type sequence to accommodate this factor (Fig. 4.1). In the case of Newcastle, the continued development of the railway network in northern New South Wales enabled the port to demonstrate its resilience and continued growth in 1911 (Fig. 4.5).

The process of lateral connection had spread rapidly and by 1911 affected Queensland, Tasmania and Western Australia (Fig. 4.5. Table II). Brisbane and Hobart were clearly benefitting from expanded hinterlands and the reorientation of the railway network on Perth (Perth–Albany railway 1889) in Western Australia enabled Fremantle, with its new harbour works, to replace Albany as the principal port of Western Australia and the major outlet for the Eastern Goldfields—Coolgardie and East Coolgardie (1892–3), Kalgoorlie (1893). Mail contracts were transferred to Fremantle and the port also deprived Albany of its function as a coaling station. Indeed, so great was Albany's eclipse that it failed to maintain its influence within the area closest to the port. Such a rapid change in the fortunes of ports emphasises the role of the railways and the influence of political patronage in making and breaking a port.

The fourth phase: centralisation

With the further extension of the railway network and the development of motor transport (to handle increased agricultural outputs stemming from the application of irrigation, dry farming, fertilisers and new machinery, and expanded imports following Confederation in 1901) the centralisation of activities on a state basis, as illustrated by P_4 in the fourth phase of the idealised type sequence, was feasible (Fig. 4.1). Even in 1911 before road transport had shown any marked impact Melbourne already dominated the trading activities of Victoria (Fig. 4.5). By 1938 the centralisation of activities was pronounced in South Australia, Western Australia, Queensland and Tasmania (Fig. 6). Despite the link-up of all of the penetration lines (with the exception of the Cooktown and Normanton lines Brisbane's off-centre location in Queensland prevented the port from depriving the northern ports of direct overseas shipping (Table II). Fremantle, linked to South Australia by the completion of the Trans-Continental Railway Line from Kalgoorlie to Port Augusta (1917, extended to Port Pirie 1937), was now dominant (Table II). In some ways, the link was illusory as the line from Fremantle to Kalgoorlie was 3 ft 6 in., while the new line was 4 ft 8½ in. Similar gauge problems in South Australia with

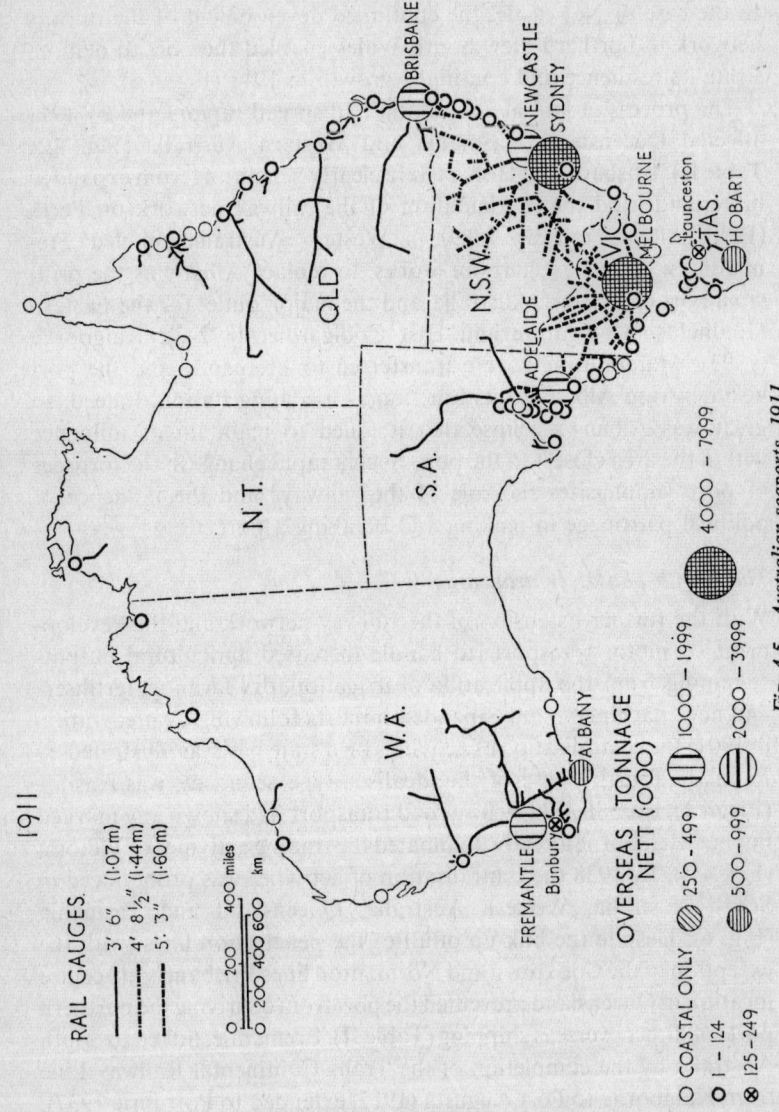

Fig. 4.5 Australian seaports, 1911

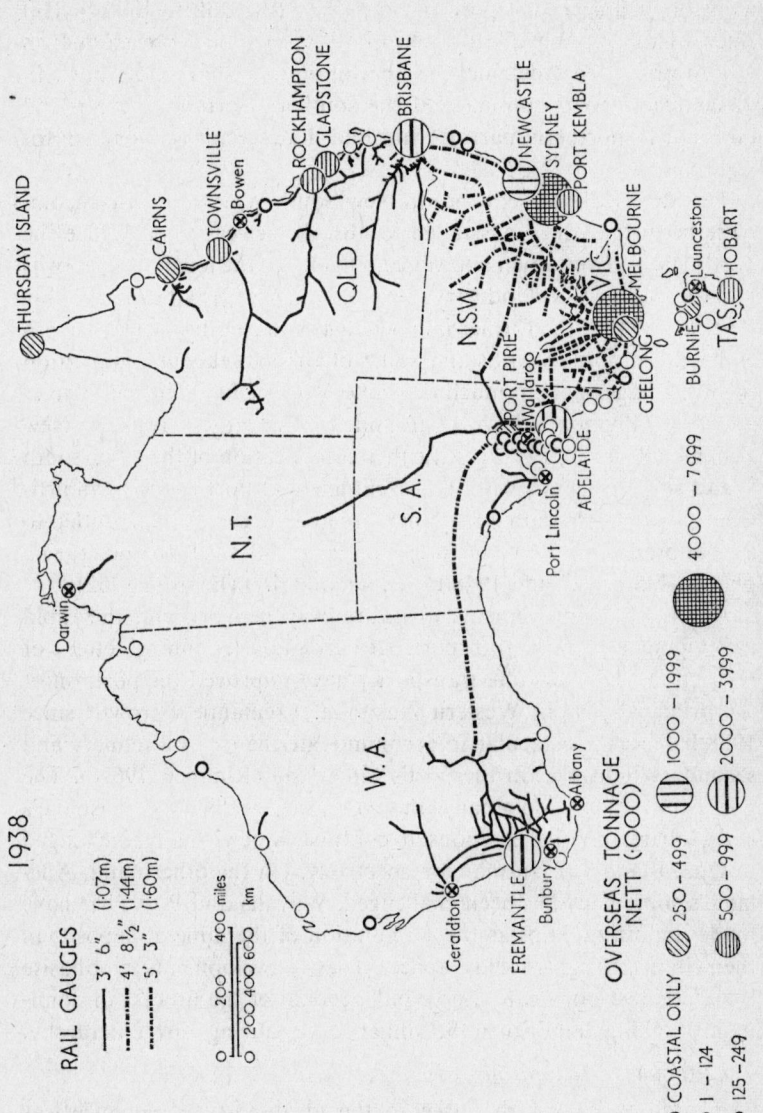

Fig. 4.6 Australian seaports, 1938

the 5 ft 3 in. and 3 ft 6 in. lines together with the marked development of such specialist ports as Port Pirie (the link to Broken Hill
since 1883) and Whyalla (developed 1909 to export ore from Iron
Knob) prevented Adelaide from becoming the State's sole port. In
Tasmania, Hobart commanded the southern portion of the island
but in the northern part Burnie and Launceston contended for
supremacy.

The idealised-type sequence in New South Wales was also incomplete because Sydney has failed to absorb Newcastle or reduce the
port's status. Even when allowance is made for the local coal deposits
and the iron and steel industry (developed since 1915) it can probably
be assumed that the former independence enjoyed by Newcastle has
enabled the port to withstand some of the subsequent competition
from the larger port in much the same way as Bluff, Timaru, Napier
and New Plymouth have confounded their larger rivals in New
Zealand (Rimmer, 1967*b*). A further complication of the New South
Wales situation came with the development of Port Kembla, an artificial port to the south of Sydney created in 1898 as an outlet to
a coal-producing area and later for iron and steel manufacturing
plants—built 1927 and 1948 (Figs. 4.6 and 4.7) (Britton, 1962).

Additional improvements in the railway network and the rapid
development of road transport, free from the confining fetters of
restrictions on inter-state transport,† have improved the position of
the principal port in Western Australia. Fremantle's growth since
1938 has been spectacular to accommodate the new oil refinery and
secondary industries in the port's hinterland (Rimmer, 1967*c*). The
port is now almost on a par with Sydney and Melbourne. Brisbane's
and Hobart's relative positions, in contrast, were virtually unchanged
in Queensland and Tasmania respectively. On the other hand, Adelaide's supremacy has been challenged. Whyalla and Port Pirie have
made important gains as the exploitation of the mineral deposits in
their hinterlands proceeds apace. These developments emphasise
that specialist ports can emerge independent of the process of dominance ranking and disturb the influences producing a port hierarchy.

The fifth phase: decentralisation

The addition of a fifth phase to the idealised-type sequence was
necessary in Australia to explain the changes in the status of ports

† From the 1930s until 1954 individual States regulated inter-state transport to
limit competition against the State Railways. A judgment given by the Privy Council in 1954 stated that such restrictions were against the Australian Constitution.

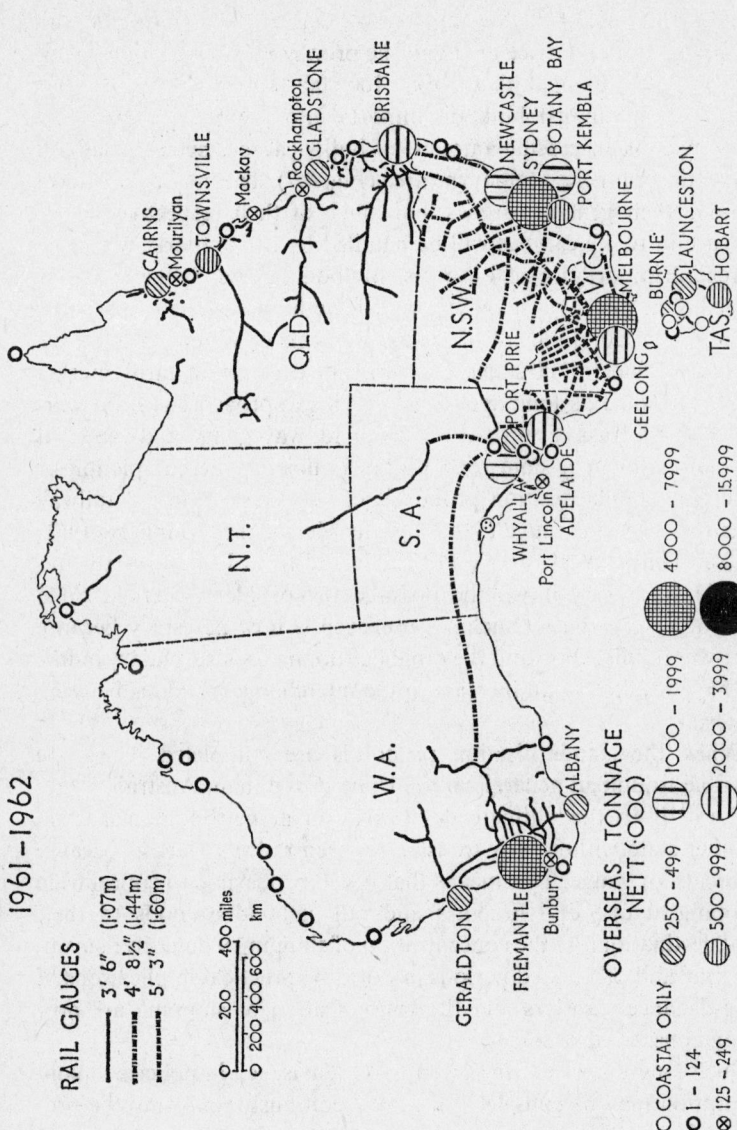

Fig. 4.7 Australian seaports, 1961/2

more fully. This need has arisen from the inability of Sydney to accommodate all of the low-value bulky cargoes being imported and exported from its service area and the problems experienced by Melbourne in handling tankers over 100,000 tons d.w.t. in 1961–2. Botany Bay handled bulk oil imports into the Sydney area and Geelong took increasing amounts of bulky cargoes such as bulk oil and fertiliscr for the Melbourne area (Fig. 4.7). The major ports have retained their general cargoes, but much of their respective State's increase in bulky cargoes will be handled by the outports. With the increased importance of outports, Melbourne and Sydney did not exceed 8 m. tons as otherwise would have been expected.

So far the development of ports has been considered within the context of individual states even though a common tariff market was established as long ago as 1901. This approach has been warranted in the past because of breaks of railway gauge and the legal restrictions of inter-state road traffic. Following the completion of all of the standardisation proposals of converting selected railway lines to 4 ft 8½ in. gauge (Wodonga to Melbourne completed 1961, Broken Hill to Adelaide via Port Pirie and from Kalgoorlie to Fremantle via Perth) it will be possible to consider Australia on a Commonwealth basis. Such an approach is now necessary because post-war immigration and the rapid development of secondary industry have resulted in an increase in the interchange of goods between States.

When the standardisation proposals are completed it will be possible to unload general cargo at one of the main Australia ports and rail it to the ultimate destination in a metropolitan area in another state without any transfer between railway wagons because of breaks of gauge. This means that it will be feasible to concentrate shipping at one or two ports and rail high-value goods to their final destination. As the concentration of shipping would save steaming time and protracted periods in port, the practical implications of long-distance transfers and the concentration of shipping are now being considered seriously.

Such long-distance transfers of goods envisaged in policies of concentration may be completed more expeditiously following the proposed introduction of containerisation. Goods will be packed in containers at their place of origin, transferred to a ship, moved by sea to the destination port and railed or trucked to the final consumer without breaking the seal of the receptacle. Insurance fees, pilferage and damage would be reduced considerably following the adoption of this principle. Already specially designed container berths are under construction at Melbourne and Sydney.

The completion of these berths and the introduction of containerisation may permit the emergence of either Sydney or Melbourne as a super port. Sydney is clearly in the more advantageous position because the completion of the standardisation proposals will give the port direct rail access to Brisbane, Adelaide, Melbourne and Fremantle. As the rail link between Melbourne and Adelaide is not to be standardised, Sydney, therefore, may capture the very high value container traffic originating in and destined for all of the mainland States except Victoria. Should Sydney emerge as the nation's sole super port it would be necessary to recast the idealised-type sequence if the model is to simulate reality.

REFERENCES

BIRD, J. (1963). *The Major Seaports of the United Kingdom*, London.

BLAINEY, G. (1966a). *The Tyranny of Distance, How Distance Shaped Australia*, Melbourne, especially pp. 70–98.

BLAINEY, G. (1966b). Ibid., pp. 228–43.

BLAINEY, G. (1966c). Ibid., pp. 118–47.

BLAINEY, G. (1966d). Ibid., p. 70.

BLAINEY, G. (1966e). Ibid., pp. 240–3.

BLAINEY, G. (1966f). Ibid., pp. 244–65.

BOXER, B. (1961). *Ocean Shipping in the Evolution of Hong Kong*, University of Chicago, Department of Geography Research Paper No. 72, Chicago.

BRITTON, J. N. H. (1962). *The Growth of Port Kemble*, Geographical Society of New South Wales and Department of Geography, University of Sydney, Research Paper No. 2, Sydney.

KENDALL, M. G. (1948). United Kingdom shipping statistics, *Journal Royal Statistical Society*, Ser. A (General) Part II, Vol. III, p. 40.

LEARMONTH, N. E. (1960a). *The Story of a Port; Portland; Victoria*, Portland.

LEARMONTH, N. E. (1906b), Ibid., p. 39.

McKENZIE, A. G. (1918). Harbour development, *Australian Town Planning Conference 1917* . . . Official Volume of the Proceedings, Adelaide, p. 101.

NEALE, E. P. (1955). *Guide to New Zealand Statistics*, 3rd edn., Auckland, p. 116.

PERRY, T. M. (1963). *Australia's First Frontier*, Melbourne.

RIMMER, P. J. (1966). The problem of comparing and classifying seaports, *The Professional Geographer*, **18** (2), 83–91.

RIMMER, P. J. (1967a). The changing status of New Zealand seaports, 1853–1960, *Annals of the Association of American Geographers*, **57**, 88–100.

RIMMER, P. J. (1967b). Ibid., pp. 95–100.

RIMMER, P. J. (1967c), Changes in the ranking of Australian seaports 1951–2/ 1962–2, *Tijdschrift voor Economische en Sociale Geografie*, **58**, 28–38.

SERLE, G. (1963). *The Golden Age—A History of the Colony of Victoria 1851–1861*, Melbourne.

SHAW, A. G. L. (1960a). *The Economic Development of Australia*, 4th edn., pp. 37–47.

SHAW, A. G. L. (1960b). Ibid., pp. 48–60.

SMITH, R. H. T. (1962). Commodity movements in southern New South Wales, Department of Geography, Australian National University, Canberra.

SMITH, R. H. T. (1964) The development and function of transport routes in southern New South Wales, *Australian Geographical Studies*, **2**, 47–65.

TAAFFE, E. J. *et al.* (1963). Transport expansion in underdeveloped countries: a comparative analysis, *Geographical Review*, **53**, 503–29. See pages 32–49 of this volume.

WEIGEND, G. G. (1956). The problem of hinterland and foreland as illustrated by the Port of Hamburg, *Economic Geography*, **32**, 1–16.

5 Transport Expansion in Liberia

WILLIAM R. STANLEY†

WITHIN the past twenty-five years, Liberia has been transformed from a series of disconnected coastal settlements to what is essentially a politically unified country noticeably throbbing with economic activity and social progress. This change can be attributed primarily to the impact of port and road development, which began toward the end of World War II. In the main, modern port construction and the development of land communications have been by-products of other stimuli, primarily the stationing of United States military forces in the country during the war, and more recently the discovery and development of high-grade iron ore deposits in the Bomi Hills, the Nimba Mountains, the Bong Hills, and along the Mano River. The presence of United States troops led directly to the construction of the deepwater port at Monrovia; iron-ore extraction made its expansion necessary and led to the construction of a more modern facility at Buchanan, 60 miles to the southeast. Recent iron-ore exploration in the Wologisi Range in northwestern Liberia may lead to the building of another iron-ore loading facility at Robertsport, near the Sierra Leone border. Greenville, the third major port in the country, was built with assistance from the German Federal Republic to facilitate the export of bananas and timber; German companies have been active in developing the export potential of both these commodities.

Expansion of transport facilities in Liberia approximates the ideal-typical sequence suggested by Taaffe, Morrill, and Gould in 1963. (See pages 32–49 of this volume). They envisioned six sequential categories of development, some of which may be present in different parts of a country at the same time. These categories are: (1) scattered ports, (2) penetration lines and port concentration, (3) development of feeders, (4) beginnings of interconnection, (5) complete interconnection, and (6) emergence of high-priority 'main streets'. The history of transport expansion in Liberia can be closely related to each of the first four steps in the ideal-typical sequence. On the other hand, at least one high-priority 'main street' is identifiable before the interconnection of inland penetration routes has been attained.

† This is a shorthand version of Professor Stanley's study published in 1970.

This study attempts to interpret changes in maritime and land transport systems and to note events peculiar to Liberia by examining three broad stages of transport expansion—scattered ports, the modern period of port construction, and the consolidation and interconnection of roads. The modern period of port construction corresponds roughly to penetration lines and port concentration in the ideal-typical sequence, and consolidation and interconnection include the development of feeder roads and the beginning of interconnection in the ideal-typical sequence.

SCATTERED PORTS

From the fifteenth century—the time of earliest contact with Europeans—until about 1950, Liberia's stage of transportation development can be classified as one of scattered ports. By the nineteenth century a score of coastal trading stations had been established in Liberia. These were usually located near the mouths of large rivers or where the curvature of the coast provided adequate anchorage for European ships (Fig. 5.1). The Dutch, from the time of their first appearance in West Africa in the seventeenth century, intermitently maintained a small trading post at Cape Mount (Robertsport), but the Grain Coast—as this stretch of coast was then called—never had the massive and more permanent trade forts and castles common to the Gold and Benin Coasts to the east (Lawrence, 1963). The absence of trading forts in Liberia was attributable not to lack of commerce but rather to more lucrative trade farther east; (see the account of Captain William Towerson's voyages to the Grain and Gold Coasts in the 1550s, quoted in Blake, 1937). Most of these early trading sites have long been disused, though a few have developed into major ports and some linger on as relic ports. Today the relic ports, usually small America-Liberian settlements or native villages of some consequence, are supported by rare visits from tramp vessels or, more commonly, by coastal steamers that transship export products collected from their small hinterlands to a larger port and bring in imported goods. Of the relic ports, only Robertsport is currently connected to Monrovia by scheduled freight and passenger coastal service. This is due primarily to Robertsport being the seat of Grand Cape Mount County and the birthplace of many prominent Liberians. There seems little possibility of further development at Robertsport unless the iron-ore deposits in the Wologisi Range are developed, and a railroad is built to link

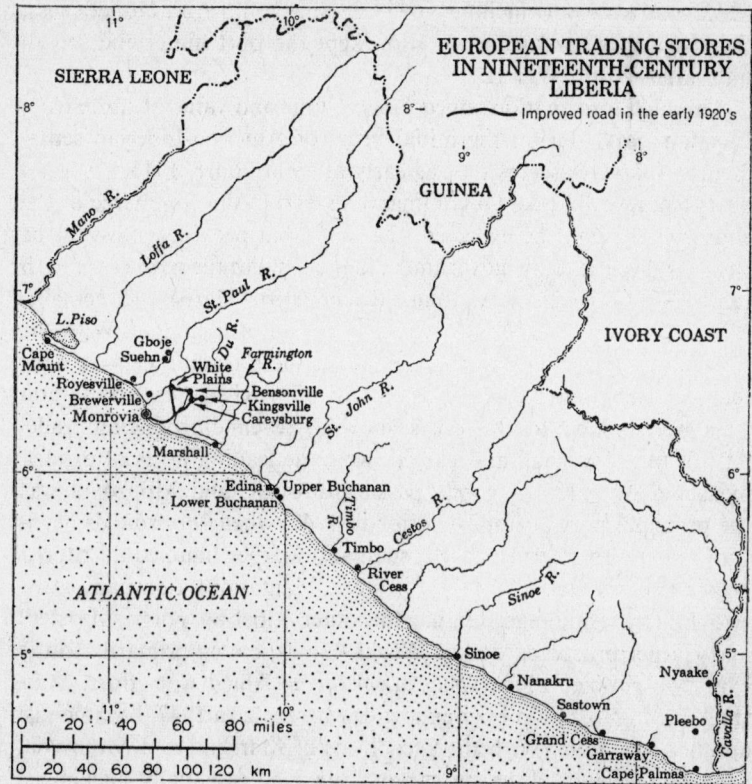

Fig. 5.1   European trading stores in nineteenth-century Liberia

them with Robertsport; agricultural exports alone can hardly justify major development of this open roadstead.

The early growth of trading stations that are now major deep-water ports was largely the result of settlement by Americo-Liberians after 1821. These settlers required imported goods and sought to maintain communication with Europe and the United States. The degree of their dependence on imports is shown by Christy's comment to the effect that during World War I Monrovians were obliged to live on the produce of the country for the first time (Christy, 1931). Subsequent growth and development of the principal port cities—Monrovia, Buchanan, Greenville, and, to a lesser extent, Harper—derived from the expansion of Americo-Liberian control over hitherto isolated parts of the country and from the need to provide port facilities for increasing commerce. Nonetheless, the absence of land transportation kept the port hinterlands small until after World War II.

Although information concerning volume and value of trade from Liberian ports during the initial years of Americo-Liberian settlement is sparse, it seems that as early as 1906 (Starr, 1913*a*), several ports had already taken a commanding lead in the commerce of the country. As might be expected, the ports that provided most of the customs revenue to the government in the first decade of the twentieth century—Monrovia, Buchanan, Robertsport, Harper, Greenville, Marshall—were also the principal immigrant settlements. With the exception of Marshall and Robertsport, these trading centres have retained their supremacy to this day. By 1939 River Cess and Grand Cess were added to the ports showing revenue-producing trade. At this time Marshall and Harper were the leading ports of export, measured by value of goods passing through, and Monrovia was the principal port of import. Prior to 1947, when Monrovia's deepwater port was completed, Marshall and Harper handled all rubber grown and purchased by the Firestone Plantations Company at its two Liberian holdings. Unlike the other Liberian ports, Marshall and Harper prospered during World War II, owing primarily to the export of rubber, which was critical to the Allied war effort. Marshall, having lost its rubber export to Monrovia in 1947, became just another relic port. With the exceptions of Marshall and Monrovia, export hinterlands of the major ports remained essentially unchanged during this period of transport development.

THE MODERN PERIOD OF PORT CONSTRUCTION

The need for a port capable of berthing ocean-going ships was apparent well before World War II. Indeed, construction of such a port was incorporated into the Liberian government concession agreement with Firestone in 1926. Firestone obligated itself to spend up to $300,000 toward construction of a port at either Monrovia or Marshall, the final choice to be determined by the availability of rock for the breakwaters. Monrovia was selected, but after spending $115,000 Firestone concluded that a harbour could not be built for less than several million dollars and received permission from the government to cancel this part of the concession agreement (Taylor, 1956).

Port building began in earnest in 1944, when the United States undertook construction of a deep-water facility at Monrovia as a result of agreements that permitted the stationing of American forces in Liberia.† Monrovia seemed a logical choice for the site of the country's first deep-water port. Not only was it the capital and largest city, but, equally important, it was at the apex of what little road mileage existed in Liberia. Near Robertsfield, a military airfield built by the United States during the war, were the Harbel estates of the Firestone Plantations Company, which were already linked by road to Monrovia. Intermittent stretches of road in the interior had been built by levies of indigenous peoples as part of the country's public works programme, and by 1943 these already formed the outline of what was to become the principal highway from Monrovia to Ganta. As a result Monrovia soon attained a pre-eminent trading position among the Americo-Liberian coastal settlements. Although as late as 1939 it ranked fourth among Liberian ports in the export of agricultural products, it took the lead by 1948 and has maintained it into the 1960s. Even if shipments of natural rubber were omitted, Monrovia would still dominate the export agricultural trade of the country.

Twelve years after the opening of Monrovia's port, the Liberian government embarked on a programme to develop the commerce of the Americo-Liberian settlements of Greenville and Harper in the southeast. Construction of a port at Greenville resulted from a 1953 compact between Liberia and the African Fruit Company, a German-owned concession, which agreed to build a port for use by banana ships. The German federal government provided Liberia with a long-term loan of $10,500,000 for this purpose. Although a breakwater and dock were constructed at Blubara Point across from Greenville before 1960, it wasn't until 1962 that storage facilities and cargo-handling equipment were installed. Banana blight, which resulted in eventual conversion to natural rubber cultivation, and the lack of access to the extensive timber reserves in the interior, caused almost total stagnation of Greenville from the late 1950s until 1963, when the first rubber was ready for export.

† 'Memorandum of Conversation, by the Assistant Chief of the Division of Near Eastern Affairs (Villard),' in *Foreign Relations of the United States*, *Diplomatic Papers, 1943*: Vol. 4, *The Near East and Africa*, U.S. Department of State Publ. 7665, Washington 1964, 660. See also 'Construction of a Port and Port Works Agreement between the United States and Liberia,' U.S. Department of State Publ. 2186, Executive Agreement Ser. 411, Washington 1944.

Limited harbour facilities at Harper have severely handicapped its economic growth and the development of its relatively rich hinterland. Firestone has a rubber plantation near Pleebo, and there is extensive timber acreage beginning at Nyaake. In 1959 a protected anchorage was developed at Harper in an area lying between Cape Palmas and Russwurm Island. However, the construction of the breakwater removed too much rock from Russwurm Island, and its profile was lowered to such an extent that it could no longer serve as an adequate barrier against the sea. In consequence, warehouses and petroleum storage facilities on the island have frequently been inundated. The poor construction of this port has made its use difficult, even by the shallow-draft coastal vessels for which it was designed. In 1964 additional funds were requested from the United States to improve navigation and port facilities here, but to date nothing has been done.

The newest deep-water port in Liberia was constructed at Buchanan by the Liberian American–Swedish Minerals Company (LAMCO). This port, which was opened to traffic late in 1963, contains some of the most modern iron-ore loading facilities in the world. A railroad 167 miles long, links the port to the ore deposits at Mount Nimba in northern Liberia. The cost of port, railroad, and mine facilities represented an overall investment of approximately $200 million. Selection of Buchanan as a port site was due primarily to the availability of rock in sufficient quantity for the construction of two breakwaters. This port has transformed a quiet coastal settlement into one of the larger cities in the country. The need to move in construction equipment led to the building of a modern road that links Buchanan to Monrovia by way of Firestone's Harbel plantation. In 1966 a concrete bridge across the St John River replaced a ferry, making Buchanan the first of the larger Americo-Liberian settlements along the coast to have direct road connections with Monrovia.

GROWTH OF THE ROAD NETWORK

Modern port construction has helped to develop a road network. By 1965 the basic outline of Liberia's road system had been established (Fig. 5.2). The separate roads that existed in 1945 have now been joined, primary roads have been extended to connect with the Sierra Leone, Guinea, and Ivory Coast road systems, and another road goes southeastward from Ganta to connect with the main road in

Maryland County which serves the Firestone plantation and the fledgling timber companies. In addition to these main roads, other routes have been built from Brewerville to Bopolu, from Sanniquellie to Kahnple, from Kakata north to the St Paul River, and from

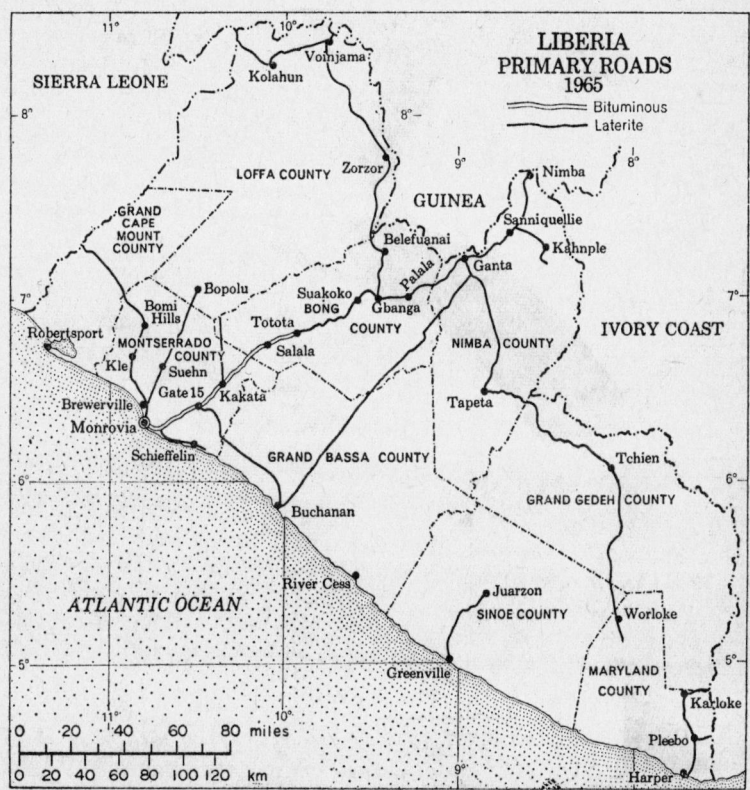

Fig. 5.2   Liberia: primary roads, 1965

Greenville to Juarzon. These routes have been further supplemented by roads constructed by private corporations to provide access to their mining sites. The private roads, built originally for company traffic, are used extensively for general traffic and have become incorporated into the national road system. Significant stretches include the road from Monrovia to Bomi Hills built by the Liberia Mining Company and later extended as far as the Mano River to serve the iron-ore mine there; LAMCO's road from Sanniquellie to the Nimba iron-ore mine; the LAMCO service road from near Ganta to Buchanan, paralleling the railway; and, finally, the road from Buchanan to Monrovia via Robertsfield.

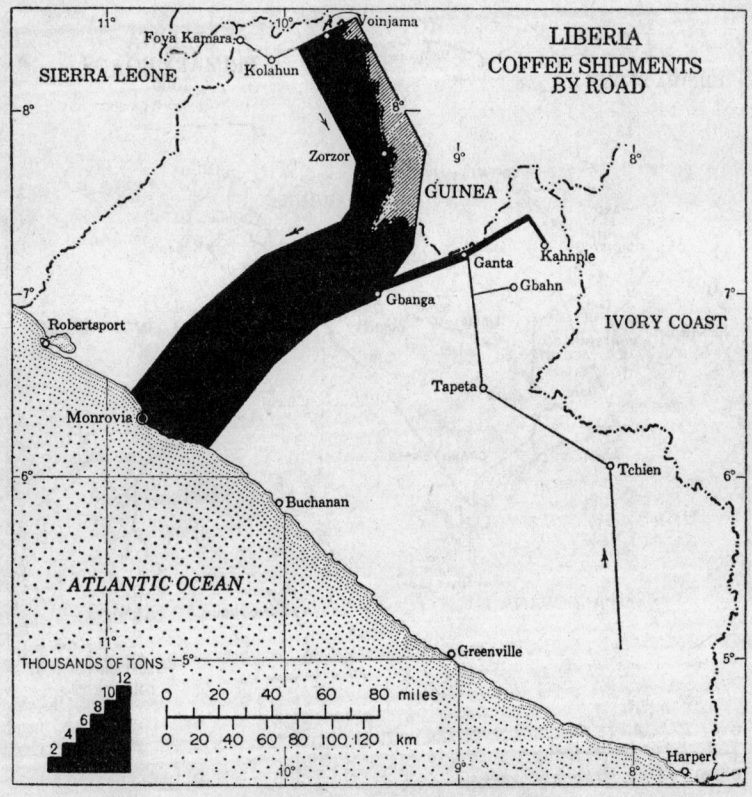

Fig. 5.3   Liberia: coffee shipments by road

These roads are unevenly distributed throughout the country, a situation that has favoured certain ports. Monrovia, for instance, has always been best served by the road network. Its hinterland encompasses the greater part of the country, and its port has attracted most of Liberia's agricultural exports. Figure 5.3 depicts the movement of coffee by road from interior collection points to Monrovia. Although other ports also export this commodity, in 1965 their hinterlands were relatively small as compared with that of the capital. The flow of coffee into Monrovia is misleading, since it does not represent Liberian production alone; a great deal of coffee grown in Guinea is brought illegally across the border near Voinjama by farmers who seek higher prices.

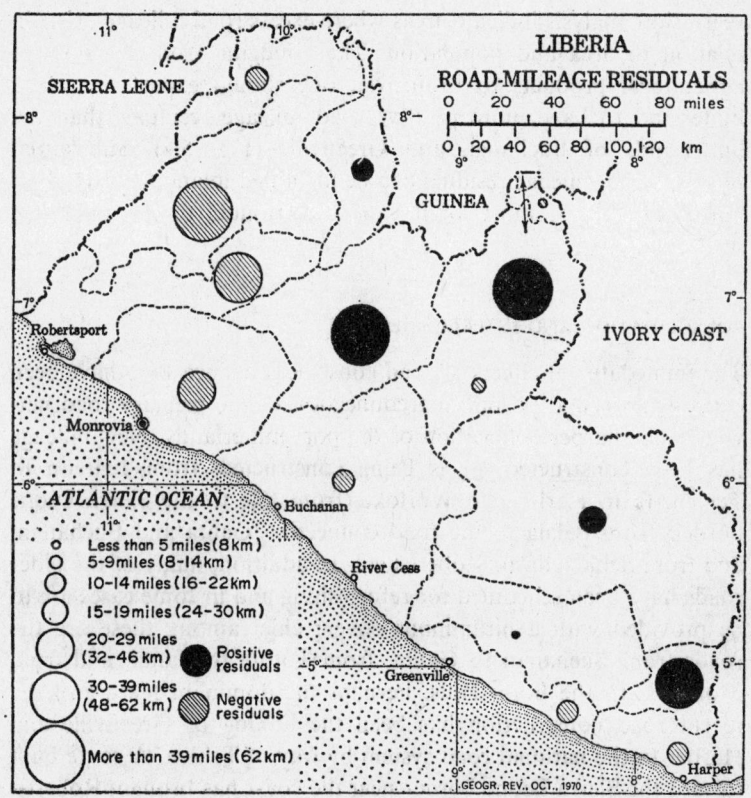

Fig. 5.4   Liberia: road mileage residuals

Monrovia's attraction for commerce moving by road can be further explained by the use of a basic regression model. In such a model, road mileage is treated as a dependent variable, population and area as independent variables. In order to increase the number of population units in the regression analysis, the fifteen population-reporting units derived by Porter were utilised (Porter, 1956). The results of the basic regression analysis suggest a close relationship between the distribution of road mileage and population. Approximately 90 per cent of the road mileage is associated with population and area,† with population accounting for more of the variation in road mileage than area accounts for in 12 of the 15 population reporting units. Road mileage residuals from the regression analysis indicate areas where excess road mileage exists in relation to area and population. Thus, judging from the flow of agricultural products to Monrovia, one should expect this city's hinterland to have more positive road mileage residuals than the hinterlands of Buchanan and Greenville (Fig. 5.4). Substantial positive road mileage residual can be identified inland from Harper, but much, if not most, of it can be attributed to the Firestone planation.

CONSOLIDATION AND INTERCONNECTION

The immedate net effects of road construction since 1965 have been consolidation and partial interconnection of the primary road network and sharper delineation of the port hinterlands. New mileage has been constructed, or is being constructed, from Juarzon to Tchien, from Karloke to Worloke, from Kle to the Sierra Leone border, from Palala to the road connecting Ganta and Buchanan, and from Schieffelin to Robertsfield. In addition, many of the older roads have been scheduled tor refurbishing and in some cases are to be provided with a bituminous cover; chief among these are the roads from Buchanan to Ganta, from Totota to Ganta, and from Monrovia to the Bomi Hills. Perhaps the dominant feature of the recent road construction has been the linking of Greenville and Harper to the national road system by way of Tchien. A route built toward the Sierra Leone border near the coast has brought Robertsport and Monrovia closer together and has permitted Monrovia to capture Robertsport's immediate hinterland. In the case of Greenville and Harper, however, new road construction has greatly expanded their hinterlands and Monrovia should no longer attract

† The regression equation is $Y_c = 0 \cdot 6424 + 0 \cdot 0104 X_1 + 0 \cdot 0009 X_2$, when Y_c is the estimated highway mileage, X_1 the area of the reporting unit, and X_2 the population of the reporting unit. The r^2, or explained variation, is $0 \cdot 8775$, or 88 per cent. Considering the relatively small population of Liberia (approximately 1,016,000 in 1962), the number of miles of road in 1965 was shown to be all the more minute when a change of population of one person would have had an effect of change of $0 \cdot 0009$ miles.

the preponderance of commerce from the southeast, particularly from Grand Gedeh County and the northern parts of Sinoe and Maryland Counties. Considering distance only, export commerce in these areas should be drawn to either Greenville or Harper rather than to Monrovia or Buchanan.

The inland extension and linking of the principal roads have not only consolidated the road network but, equally important, have led to the initial stages of interconnection through the development of feeder roads. Although the construction of feeder roads has proceeded most rapidly in Monrovia's hinterland, they are now also being built inland from other ports, particularly in the southeast.

The beginning of road interconnections can be depicted graphically. Figures 5.5, 5.6 and 5.7 present the extent of road development

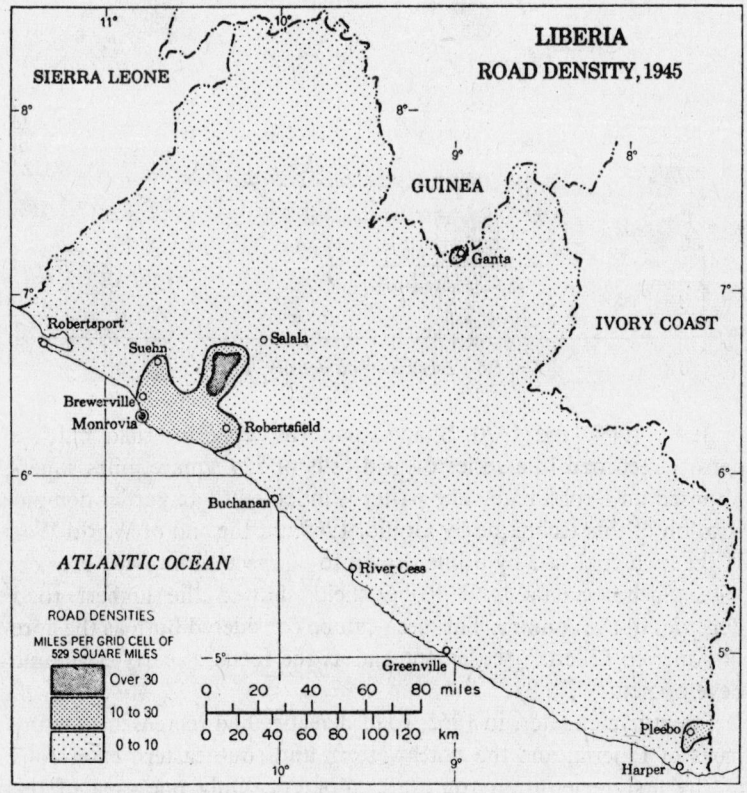

Fig. 5.5 Liberia: road density, 1945

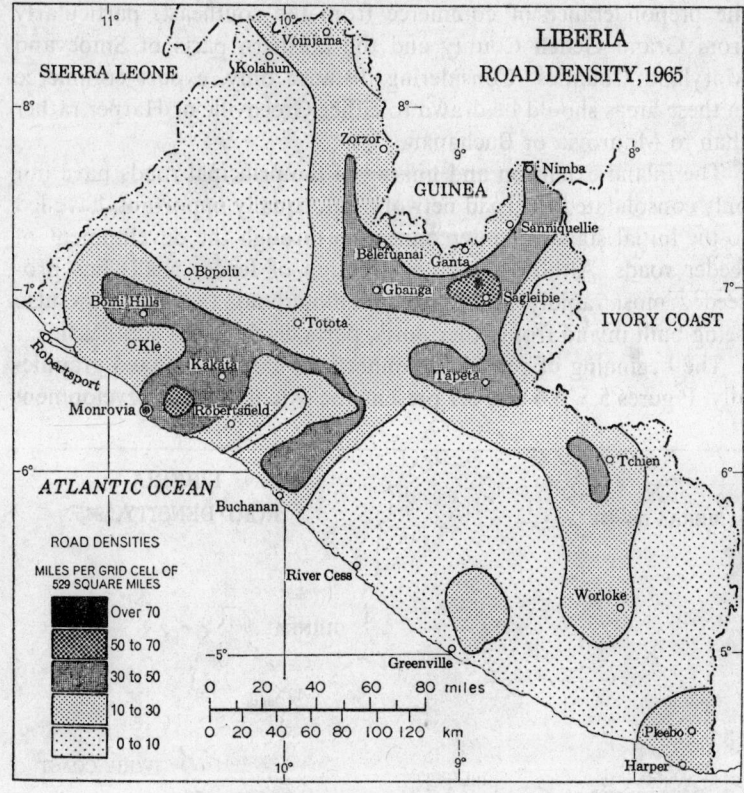

Fig. 5.6   Liberia: road density, 1965

in 1945, 1965, and 1970. Shaded patterns represent road mileage density recorded in a series of grid cells of 529 square miles superimposed on a road map. The writer is indebted to an earlier demonstration of this technique by Gould, 1960. At the end of World War II the area enclosed by a line drawn to connect Monrovia, Brewerville, Suehn, Salala, and Robertsfield showed the highest road densities in the country. This area can be considered both as the core of America-Liberian settlement and as the focus of early economic development.

Twenty years later, in 1965, road densities had increased in many parts of Liberia, and the northwestern and southeastern extensions of the first penetration route are apparent. Only one area of the highest density category (more than 70 miles of road per grid cell) was in evidence at this time—between Ganta and Sagleipie, near the point where the Ganta–Tchien and the Ganta–Buchanan roads intersect. Areas of lesser density (50–70 miles per grid cell) can be seen inland from Monrovia and surrounding the high-density area of Sagleipie. Road density had expanded in the direction of Bomi Hills and the Sierra Leone border, as the result of iron-ore mining and subsequent road building at the Bomi Hills and Mano River mines. There remained a distinct gap in density between Monrovia and Buchanan. This gap is further evidence that road construction in Liberia has progressed inland from the several coastal settlements rather than along the coast to link the settlements. Construction near the coast is, as a rule, more costly than in the interior, owing to the larger number and greater length of the bridges required to cross

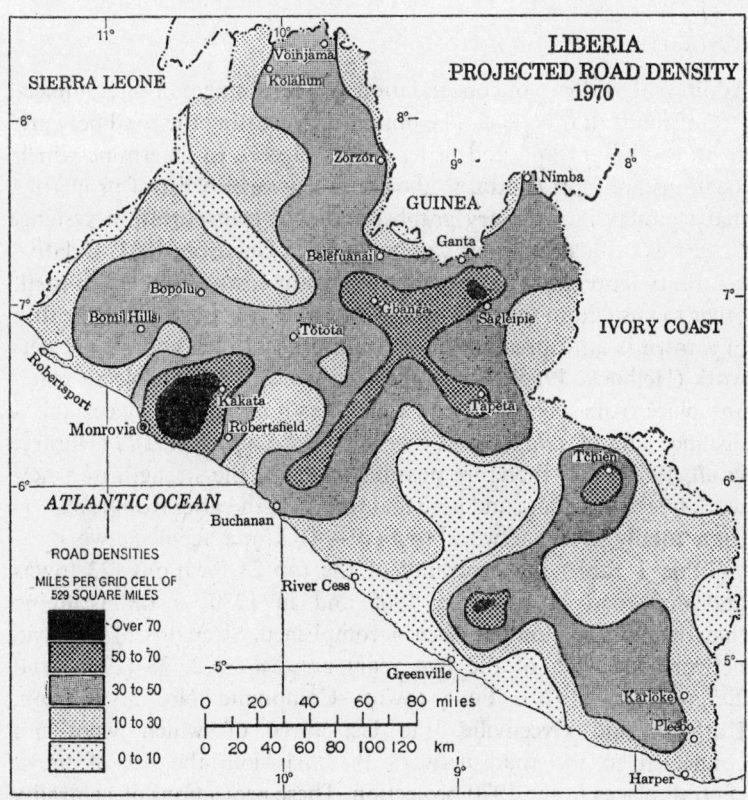

Fig. 5.7 Liberia: road density, 1970

the rivers. Some improvement of accessibility by road had also taken place in the Harper–Pleebo vicinity and inland from Greenville.

If road construction planned for the period ending in 1970 is completed, three areas will have road densities greater than 70 miles in each grid cell (Fig. 5.7). In addition to Sagleipie and Monrovia, the area north of Greenville should also have a high density. Construction of new roads will bring most of the eastern quarter of the country within at least the 10–30-mile category. There will still be several areas of lowest road density (0–10 miles per grid cell), particularly between River Cess and Greenville, and in the west central parts of the country, especially along the Mano River. Nevertheless, the overall effect of the planned construction is to increase and equalise road densities throughout Liberia.

CENTRALITY

Additional evidence of consolidation and early stages of interconnection of roads in Liberia can be obtained by treating the road network as an ordinary graph and using matrix algebra to determine which locations are most central to the entire network. A list of properties that characterise ordinary graphs, and thus transportation systems treated as ordinary graphs, can be found in Garrison, 1961. Relative centrality represents the degree to which any town is connected to all other towns within the network. Alternatively, the relative centrality of a town is an expression of its 'centralness' to the entire road network (Helbock, 1965). For a graph in which it is possible to reach any place from any other, and which has p number of places and a distance sum of Σd_{ij}, then $p(p-1) \geq \Sigma d_{ij} \geq p^2(p-1)/2$ (Harary, et al., 1965). It is therefore possible to define the strength of a network by the term Σd_{ij} and note that the strength varies from $p(p-1)$ for a 'totally strong network' to $p^2(p-1)/2$ for a 'feeble network'.

Table I ranks the centrality of the top 24 locations (23 towns and one road junction) in 1965 and in 1970, again assuming that the construction has been accomplished. Strengths of the road network for each of the two years are $\Sigma d_{ij} = 2,623$ (1965) and $\Sigma d_{ij} = 3,584$ (1970). Four towns—Compound No. 3, Juarzon, Karloke, and Greenville, the last three of which were not connected to the road network in 1965—join the list of 'most central' places in the 1970 projection. These projections of centrality emphasise the increasing level of road development in the southeast

Table I. Centrality to the Road Network of Twenty-four Locations

1965			1970		
Location	Ranking	Centrality value	Location	Ranking	Centrality value
Buchanan	1	33·6	Compound No. 3	1	40·3
Ganta	1	33·6	Ganta	2	38·1
Robertsfield	3	32·4	Buchanan	3	37·8
Gate 15	4	31·2	Robertsfield	4	36·2
Gbanga	5	30·9	Sagleipie	5	34·8
Sagleipie	6	30·5	Monrovia	6	33·8
Suakoko	7	27·9	Gbanga	6	33·8
Kakata	8	27·6	Juarzon	8	33·2
Monrovia	9	27·3	Gate 15	9	32·6
Totota	10	27·0	Tapeta	10	32·3
Sanniquellie	11	26·8	Belefuanai	11	30·6
Salala	12	26·2	Suakoko	11	30·6
Belefuanai	13	26·0	Brewerville	13	30·4
Tapeta	14	24·1	Sanniquellie	14	30·1
Brewerville	15	23·8	Kakata	15	28·7
Zorzor	16	22·0	Totota	16	28·0
Nimba	17	21·9	Zorzor	16	28·0
Tchien	18	20·5	Tchien	18	27·6
Kle	19	20·2	Karloke	19	27·4
Suehn	19	20·2	Salala	20	26·9
Voinjama	21	18·9	Kle	21	26·7
Bopolu	22	17·3	Greenville	22	26·5
Bomi Hills	22	17·3	Voinjama	23	25·8
Kolahun	24	16·3	Suehn	24	25·1

and suggest that this part of Liberia is no longer isolated from the rest of the country. They further suggest that the ports of Greenville and Harper should benefit from rapidly developing primary and feeder roads in Grand Gedeh, Sinoe, and Maryland Counties.

THE ROLE OF RAILROADS

Railroads have played no role as passenger and general freight carriers. Because Liberia lacked a European colonial background, it did not obtain a transport infrastructure as early as its West African neighbours did. This precluded railroads as the first penetration lines. The four iron-ore-hauling railroads were built after World War II, but none serves as a common carrier. Treatment of the four rail lines in Liberia has been minimised in this paper, since they do not serve the general public. Moreover, their possible future inclusion in the public transport system would present serious problems of interchangeability of rolling stock, since the Bomi Hills line and its extension to the Mano River has a gauge of 3 ft 6 in., whereas the Monrovia–Bong Hills and the Buchanan–Nimba lines are the standard gauge of 4 ft 8½ in. However, the service roads that were built to facilitate the laying of track have helped to open the interior and have been incorporated into the national road network.

This is not to suggest that freight and passenger railroads have never been considered; for Liberia has a long history of railroad plans. As early as 1912 the Liberian legislature granted a concession to a German firm to construct a narrow-gauge railroad from Monrovia to Bopolu and another inland from Harper (Starr, 1913*b*). Nothing came of these schemes, or of a suggestion by Firestone in the 1920s to link their new Harbel plantation to Monrovia by a narrow-gauge line. The most recently proposed railroad scheme was a line from Monrovia to Guinea recommended in 1947 by members of a United States Economic Mission (Keim, 1947). This proposal envisioned a railroad connecting Monrovia with Zorzor by way of Suehn and Bopolu, with the possibility of an extension into eastern Guinea. It soon became apparent, however, that the value of any commerce which might be generated by such a line would be infinitesimal in relation to the cost of construction. Perhaps more important was the fact that by the time this proposal was outlined, road communications were becoming established as the principal inland transport arteries.

Port concentration and growth have led to major penetration routes into the interior, beginning with Monrovia, followed by Buchanan, and ending with Greenville and Harper. The Monrovia penetration route has generated the greatest number of feeder roads and has become the first high-priority 'main street'. This sequence of transport development seems to support the comment of Taaffe, Morrill, and Gould that 'later phases [of transportation development] typically evolve around the penetration lines' (Taaffe *et al.*, 1963). In Liberia the beginnings of road interconnection had to wait until more modern port facilities were available, and this suggests the critical role of ports in the development of inland transport. The later stages of transport expansion in Liberia indicate improved communications in the long-dormant southeast. Here the exploitation of the extensive stands of marketable timber should justify government expenditures to better the transport infrastructure.

REFERENCES

BLAKE, J. W. (1937). *European Beginnings in West Africa, 1454–1578*, London, New York and elsewhere, pp. 153–4.

CHRISTY, C. (1931). Liberia in 1930, *Geogr. Journ.*, 77, 515–40 .

GARRISON, W. L. (1961). Connectivity of the interstate highway system, *Papers and Proc. Regional Sci. Assn.*, vol. 6 (6th Annual Meeting, 1960), Philadelphia, pp. 121–37.

GOULD, P. R. (1960). The development of the transportation pattern in Ghana, *Northwestern Univ. Studies in Geogr. No. 5*, Evanston, Ill.

HARARY, F. *et al.* (1965). *Structural Models; An Introduction to the Theory of Directed Graphs*, New York, p. 187.

HELBOCK, R. W. (1965). Transportation network structure in Oceania (unpublished paper, Department of Geography, University of Pittsburgh, July), p. 16.

KEIM, P. F. (1947). Report of the reconnaissance of the Western Province of Liberia and adjacent areas in French Guinea for the purpose of determining a route for the construction of a railroad (unpublished report submitted to United States Economic Mission, Liberia).

LAWRENCE, A. W. (1963). *Trade Castleand Forts of West Africa*, London, p. 41.

PORTER P. W. (1956). Population distribution and land use in Liberia (unpublished, Ph.D. dissertation, London School of Economics and Political Science, University of London).

STARR, F. (1913*a*). *Liberia*, Chicago, p. 143.

STARR, F. (1913*b*). Ibid., p. 141.

TAAFFE, E. J. *et al.* (1963). *Geogr. Rev.*, **53**, 506. See pages 32–49 of this volume.

TAYLOR, W. C. (1956). *The Firestone Operations in Liberia*, Washington, p. 15.

6 Highway Improvements and Agricultural Production: An Argentine Case Study

FRED MILLER

HISTORICALLY, investments in highways in less developed countries have been high-priority expenditures. Often, by faith, a highway is expected to act as a catalyst for economic development. This faith is partially attributable to positive effects which have arisen from highway construction and partially due to optimistic preconstruction evaluations with forecasts which may or may not have materialised. Frequently, political leaders have been more receptive to highway projects than to other types of investments, both because highways have enjoyed favourable reputations and because they have long 'pay-off periods' and are politically 'safe'. That is, it is not known for a number of years if a given investment is worthwhile and, in the meantime, something obviously has been done.

Regardless of the rationale behind road construction, impressive results often have been achieved. Since there have been many opportunities available, sizeable benefits have accrued to societies after highway investments, even when the projects were not precisely evaluated or when alternatives were not considered. Problems are arising, however, because of the changing nature of the transportation structure in some countries and deficiencies in the current methods of appraising projects. First of all, many studies have been concerned with 'opening-up' situations in which a highway opens new territory for agriculture or, less commonly, for another industry. Now that more roads have been constructed, a more relevant question in some areas is becoming, 'Should highways be improved from dirt to pavement?' It is by no means certain that road *improvements* possess the catalytic quality that is often attributed to *new* road construction.

Second, there are some serious difficulties with highway evaluation methods. Accuracy in calculations has not been essential in examining obviously worthwhile highways, but is important when viewing road improvements. Improvements must be evaluated more rigorously since their desirability is not immediately apparent.

Appraisals have been more optimistic than might be justified since representatives of countries which are seeking funds or engineers looking for contracts have often been charged with making them. It is difficult to believe that projects are as attractive as described. In fact, seldom can an unfavourable proposal be found.

At least as important as vested interests in producing optimism is the methodology which is used. There are growing disagreements as to which benefits and costs should be included in the analyses, confusion regarding some of the direct relationships between roads and economic benefits, and, consequently, doubts about the meaning of the results. Usually, both direct and indirect benefits are considered while costs are frequently limited to 'hard core' construction and maintenance expenditures with perhaps a verbal recognition of the 'unquantifiable' factors involved. To achieve consistency, nebulous costs and benefits should either be examined in a comparable manner or a study should be made considering concrete variables to see what tangible economic effects rise from highway improvements.

This study was undertaken in an attempt to estimate the increase in agricultural production which could be gained from highway improvements†. It was especially appropriate for Argentina since there are many dirt roads, but few that are improved. Also, agriculture is easily the most important industry in the region of interest. If highway improvements were going to produce favourable results, as expected by many in the area, certainly the effects would be reflected in increased agricultural production.

No new method of evaluation was utilised in the study. Rather, there was a change in emphasis. Decisions as to whether road construction should be undertaken were not made since the totality of the benefits and costs were not considered, but they can follow from this type of study if agricultural production is considered sufficiently important to the economy. It is not denied that indirect and noneconomic benefits exist or that they can be important. It is believed, however, that direct relationships are obscured when less tangible factors are included and that more significant statements can be made if a definite influence of highway improvements on agriculture is shown. The indirect or unquantifiable benefits and

† The research for this study was done in Argentina in collaboration with John M. Hunter through a grant to him by the Agricultural Development Council, Inc., and the assistance of both is gratefully acknowledged.

costs can be weighed with greater understanding following such a study since there will be an awareness of receptivity to new ideas and of propensities to change, but these subjective aspects of the evaluation were not dealt with here.

A surprising lack of benefits was found. The reasons for this will be discussed in the text to follow. While the situation in southern Argentina is not exactly like that of many other locations, some of the explanations for a lack of positive effects are certainly applicable to different areas, and lessons for highway investments can be learned from the experience.

AREA FOR STUDY

The area of interest was the semi-arid region along National Highway 35 between the cities of Bahia Blanca in the province of Buenos Aires and Santa Rosa, capital of the province of La Pampa (Fig. 6.1). This is an agricultural zone which produces primarily cattle, sheep, and small grains, of which the majority is shipped to Bahia Blanca. Wheat accounts for about one-half the value of the output of the area and cattle comprise approximately one-quarter of the total product. While the area is generally fertile, the increase in production has been about equal to the national average which, according to the Ministry of Agriculture statistics, has increased only 0·5 per cent per year between the periods 1935–9 and 1960–3. Many of the people living in the zone feel that paving Route 35 is essential for economic progress.

The *partido* (county) Puan was chosen for specific concentration because it contains an unpaved section of Route 35 and has representative characteristics which allow the generalisation of conclusions to the region. Systematic personal contact with the producers was made possible by the focus of interest on a smaller area. The *partido* was divided into sixty-eight sections, and a farmer from each was interviewed. A questionnaire was used to gain information about farm methods and producers' aspirations as well as about expectations of production changes following road improvements. Another questionnaire, delivered to the principal marketing organisations of the *partido*, yielded data about marketing and transportation and the types of changes in them which are possible. With a knowledge of the farmers' attitudes and techniques and of the savings in marketing and transport costs, the prospective gains from road improvements were estimated. These results were then generalised to the entire Bahia Blanca–Santa Rosa region.

TRANSPORT FACILITIES

The transportation facilities in Puan are representative of those in the Bahia Blanca–Santa Rosa region. There are roads ranging from

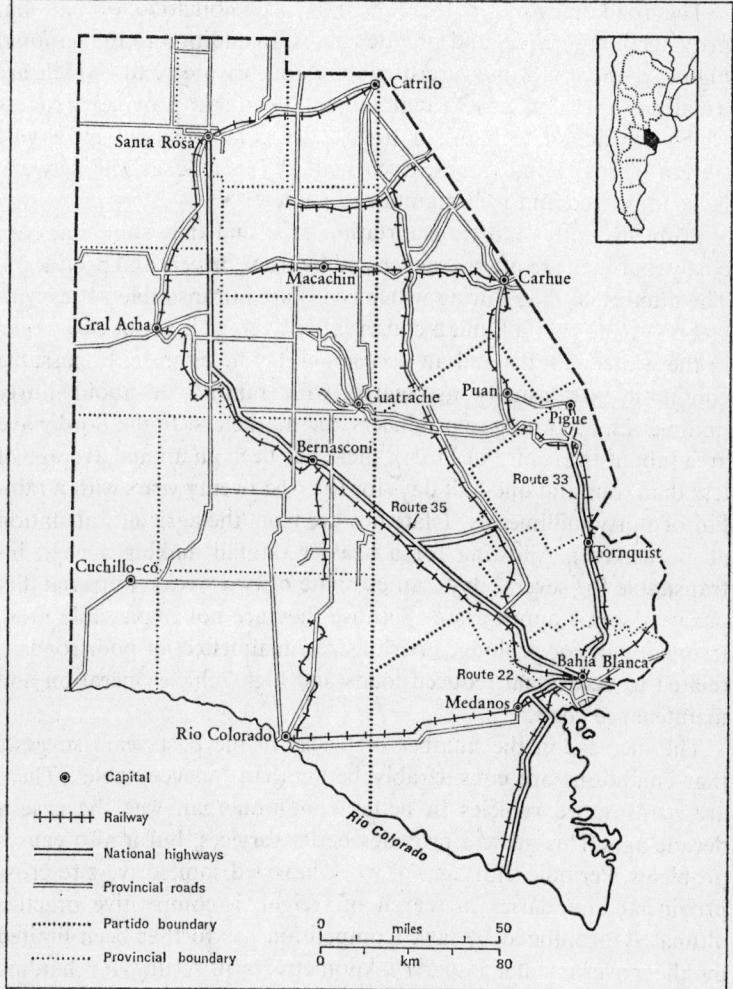

*Fig. 6.1   Transportation networks and political divisions in the Bahia Blanca–
Santa Rosa region*

overgrown ruts to newly paved highways, a noticeable increase in the size of the trucking industry, and a railroad with old cars and generally poor service. While there exists an awareness of road inadequacies everywhere in the zone, it is especially prominent in Puan because of its lack of paved highways. This is accentuated by the unpaved section of Route 35, since those living close to it are more conscious of the inconveniences which it imposes.

The road network in the region is a combination of national provincial, municipal, and private routes. In addition to the national highway, there is a mixture of public and private roads which are usually of dirt and poorly maintained. The main provincial roads, which are adequate between towns, are complemented by a vast system of roads and tracks. Even with all this mileage, the network lacks improved and well-maintained routes.

Problems with road transportation exist and they sometimes are costly, but they are often exaggerated by truck drivers and producers. The number of days during which roads are intransitable varies with soil types and rainfall, but a conservative average is about two weeks in the winter months and an occasional day in summer. Impassable conditions are usually associated with rainfall of about thirty millimetres, according to producers and marketers. In the Bordenave area (about the centre of Puan), there has been an annual average of less than four and one-half days for the past twenty years with a rainfall of thirty millimetres; (statistics are from the agricultural station at Bordenave). Allowing for a heavier rainfall making a road intransitable for several days, an estimate of two weeks' impassability per year seems appropriate. Because they are not impassable more frequently or for prolonged periods, the main effect of poor roads is related to occasional reduced loads and high vehicle operation and maintenance costs.

The increase in the number of trucks in the past years suggests that conditions are considerably better than 'unacceptable'. There are many more vehicles in better condition than was the case a decade ago. This growth provides better services, but it also causes problems. Periodic shortages of work have led some drivers to cross provincial boundaries in search of freight, a competitive practice ultimately prohibited by law. Competition has further been limited by the government's kilogram–kilometre tariff setting for hauling production anywhere within the province. The reduced mobility and fixed hauling rates have limited competition, but also have regularised service and provide more opportunities to send freight by truck.

The fixed rate per kilogram–kilometre is important since it means that a driver on the best of paved highways and worst of dirt roads receives (officially) the same payment. This situation limits the benefits which farmers can gain from road improvements. The paving of all roads might reduce truckers' costs, but allow producers only to ship more securely and rapidly without rate reduction. In practice, however, the official rate has become more a figure around which to bargain than an accepted price. On paved roads or dirt roads of good quality the official rate usually prevails, but truckers become scarce in some zones, for example the forested and sandy areas of the south, when they are not induced by a premium rate for their travel on poor roads. In other cases, when competition between truck drivers is greater or when an attempt is made to procure some of the business which otherwise would go to railroads, drivers accept a lower tariff. In effect, shipments are made on the basis of personal contracts, some of which vary appreciably from the official rate.

The informal and sometimes illegal contracts make it difficult to estimate the prospective savings for producers arising from road improvements. Generally, little change in farm transportation costs will arise from paving national or major provincial highways since, for dirt roads, they are in good condition and producers would pay an identical sum for sending goods across them after their paving. Improvements in roads in more out-of-the-way areas seemingly would lead to a reduction in transport payments, but, even in these instances, the matter is not entirely clear. The truck drivers would benefit by the road improvement as they would be able to provide more rapid services with fewer chances of breakdowns and damages to goods. With fixed rates or with little competition between truckers, however, the gains need not reach producers.

The railroads compete in varying degrees with highway transportation. When there is a large divergence in prices for their services, trucks and trains haul different products, but when the differential narrows, the differences in the qualities of the services offered cause shippers of each class of goods to decide between them. The greater speed, mobility, and reliability of trucks, and personal contact with the truck drivers cause some to utilise highway transportation, but the presence of all-weather hauling and, most of all, lower freight charges work in favour of the railways. When competition occurs, the parallel construction of the principal roads and railways intensifies it, because the points of origin and destination are practically always the same for each. Unfortunately for the railways, just as there is a frame of mind that advocates road improvement, there exists a prevailing thought that rail service is poor. Practically everyone has a story about delays, strikes, misplaced wool, or dead animals resulting from shipment by rail. Even if these instances are not common, the stories are influential if offsetting the rate advantage of the railways and in determining choices between road and rail use.

While railroads have a number of difficulties, their major problem is a lack of freight cars. Of the agricultural products in the area, railways are used primarily for shipping wheat. Trucking enterprises can compete for wheat haulage, but, with present service, it is difficult for railroads to compete for truck cargo, that is, cattle and sheep. There is much more freight which could be sent by train if the railroads' lower rates were complemented by more cars and better service. Hence, if roads were improved, it would be possible for a simultaneous change in railroads to offset any relative advantage which might have accrued to highway transportation. When the users of the transport facilities are sensitive to changes in roads and railroads, as is the case in Puan, the effect of progress by either industry must be viewed in relation to the present condition and probable future position of the other.

The future traffic flow depends on the relative improvements in roads and railways and the subsequent rate differentials. The railroads, being national enterprises, are more likely to make large changes at one time, while changes in trucking are slower because of the numerous independent enterprises. On the other hand, even with new cars and better facilities, there is no guarantee that services by the railroads will change appreciably, while with improved highways it is certain that truckers will take the easiest route.

The marketing structure within Puan is particularly suited to making the producer independent of changes in transportation facilities. The towns, because they have developed around railroad stations and, later, have become trucking centres, are the loci for organisations which handle the sale of farm products. Animals and grains are usually sold in the towns to agricultural cooperatives or independent commercial organisations. The farmers' opportunities

to gain from improved highways are limited by their current marketing methods and means of sending production to the port. Producers
make direct payments to truckers primarily for short farm-to-town
shipments. Usually, marketing intermediaries are responsible for
additional shipping so that farmers have less personal contact
with truck drivers. Consequently, if producers are to benefit, gains
from road improvements must first be received by trucking enterprises, then passed to marketing intermediaries, and finally transferred to farmers. When savings accrue to trucking enterprises, there
is no guarantee that they will be passed on to either producers or
intermediaries, or, alternatively, that the intermediaries will pass
benefits on to the farmers.

It is instructive to estimate the magnitude of cost reductions
arising from road improvements so that if government action were
to lower the truckings rates or to allow truckers to compete freely,
the savings which could be passed on will be known. A list of costs
which is published in revised form each year by the National Grain
Board allows a demonstration of the effects on wheat marketing
which can be expected to follow from road improvements (Table I).
National Grain Board data were fitted to the conditions in the area.

Table I. Cost of Production in 1966 Pesos for a
Hectare of Wheat in Bordenave

Type of cost	Cost	Percent-age of total cost	Percent-age of sales price
Direct costs			
Including plowing, seeds, herbicides, harvesting, imputed return to land, and other items	2,918	39·2	29·2
Indirect costs			
Including imputed personal income, taxes, amortisation of capital, and other items	3,327	44·7	33·3
Commercialisation and taxes exclud-ing transportation	332	4·5	3·3
Transport from farm to town	195	2·6	2·0
Transport from town to port	680	9·1	6·8
Total	7,452	100·1	74·6

Source: National Grain Board data fitted to the Bordenave area.

The figures listed are for a hectare of wheat with an assumed output of 1,000 kilograms per hectare, slightly higher than the 960 kilogram average. This farmer is assumed to live fifteen kilometres from town, a realistic assumption. As is the custom, he harvests the crop himself instead of hiring a contractor and sends it to town by truck and to the port by train (often via an intermediary). Since these data are from 1966 the actual money values have changed slightly, but the relative weights should not have been altered significantly. The elaboration of costs shows 11·7 per cent of the total cost of planting, harvesting, and marketing a hectare of wheat attributable to transportation and, of that, only 2·6 per cent is assigned to farm-to-town costs. With the present system of bonus rates for poor roads it is this 2·7 per cent which could be reduced if all roads were improved. The reduction is unlikely, since remote 'feeder roads' would not be improved before national and provincial highways. The possibility is even more improbable when it is considered that the unpaved portion of National Highway 35, if paved according to a contract being negotiated, will take at least ten years to complete from the first pavement plans.

A similar calculation is difficult to make for cattle production because of the number of reasons for which the animals are sold and variable time periods for which they are raised. While wheat is more or less a uniform product, cattle can be marketed as several distinct products, with different sexes and ages being sold for reproduction, fattening, and slaughter. The easiest and most accurate estimate of the share of the cost of transportation in the marketing process is the percentage of the sales prices which transportation comprises (Table II). For the major classes of cattle which are marketed, transportation is only from 2·6 per cent to 3·4 per cent

Table II. Transportation Costs as Percentages of Sales Prices
for Cattle Sent from Darragueira to Bahia Blanca

Type of cattle	Kilos per animal	Pesos per kilo	Value per animal	Animals per truck	Value per truck	Cost per truck	Costs as percentage of sales prices
Fat steer	400	60	24,000	28	672,000	18,805	2·8
Steer calves	220	63	13,860	45	723,700	18,805	2·6
Cows	400	50	20,000	28	560,000	18,805	3·4
Heifers	300	60	18,000	32	576,000	18,805	3·3

Note: Cattle are sent from the area in approximately the following percentages: fat steers, 10; steer calves, 30; cows and heifers, 60.
Source: Emilio M. Biondini S.R.L. (cattle marketing intermediary).

of the sales price. It is estimated that the profit margin is as high as 50 per cent for the most efficient producers. The profit-margin estimate was made by the manager of the major cattle marketing intermediary and a technician at the agricultural research station. This implies that payments for transportation are, at most, less than 7 per cent of the cost of production and marketing. This is certainly a small percentage for road improvements to effect.

Sheep and wool prices have similarly small transportation components (Table III). Transportation costs range from 8·7 to 12·3

Table III. Transportation Costs as Percentages
of Sales Prices for Sheep and Wool

Product	Basic unit	Pesos per unit	Units per truck	Value per truck	Cost per truck	Transportation costs as percentages of sales prices
Sheep						
Ewe	1 animal	850	190	161,500	19,805	12·3
Castrate Lamb	1 animal	1,200	190	228,000	19,805	8·7
Ram Lamb	1 animal	950	220	209,000	19,805	9·5
Ewe Lamb	1 animal	1,000	220	220,000	19,805	9·0
Wool						
Fine	10 kilos	1,000	1,500	2,400,000	30,000	1·3
Coarse	10 kilos	1,000	1,500	1,500,000	30,000	2·0

Notes: (1) These figures are averages of variable prices and qualities.
　　　 (2) Sheep data are for shipments from Darragueira to Bahia Blanca.
　　　 (3) Wool data are for shipments from Villa Iris to Bahia Blanca.
Source: Emilio M. Biondini S.R.L. (Marketing intermediary).

per cent of prices for the principal classes of sheep while the percentages for wool are only 1·3 and 2·0. The actual costs of production for sheep and wool are extremely difficult to calculate because of the different methods used in raising the animals and the existence of many types of wool as by-products. Transportation costs are insignificant percentages of the total costs for shipments of wool, but are more important in the case of sheep. The relatively larger share of the sales price which is attributed to the transport of sheep is because of a low price per kilo for the animals rather than high costs for trucking services. Truck rates have been increasing in the past years while the price for sheep is lower than in 1965. Even if transportation costs were reduced considerably, sheep production would not be much more lucrative because of other factors tending to depress the mutton-wool market.

The data presented for livestock and wool are only approxima-
tions. They represent averages in June 1967 and are subject to
change because of the variable prices and qualities of the products.
Nevertheless, it is not likely that they will be altered enough to
make the transportation costs as percentages of sales prices substantially different. With transport comprising 8·8 per cent of the
sales prices for wheat; cattle and wool having a much smaller
proportion, between 1·3 and 3·4 per cent; and sheep a slightly
higher figure of up to 12·3 per cent, an estimate of about 8 per cent
is a reasonable approximation of the upper limit for agricultural
products in the *partido*. With grains accounting for 50 to 60 per cent
of the value of production and cattle for 20 to 25 per cent, sheep and
wool are comparatively insignificant. Consequently, the relatively
higher transport costs for sheep and lower costs for wool do not
strongly affect the 8 per cent approximation. If the profit margin is
assumed to be 25 per cent (a figure suggested by marketing intermediaries, agricultural technicians, and available statistics), this
means that transport costs comprise about 10 per cent of the costs of
producing and marketing agricultural products. Consequently,
the monetary benefits which the truck drivers pass on to the producers must fall within a range from 0 to 10 per cent of the costs of
production. The unpaved section of Route 35 comprises just over
one-fourth the journey from Bahia Blanca to Darragueira, the
most distant marketing centre in Puan. The improvement could not
have a large effect on the total costs of shipments, especially when it
is recalled that tariffs are 'administered' prices and that any ultimate
effects depend on their being changed and these benefits accruing to
the farmers.

These computations have been made assuming that the present
conditions will be in effect when the road improvements are made
and this seems to be the best assumption. However, it is informative
to postulate a competitive trucking industry which would pass on
gains from road improvements to the producers, or to the marketing
intermediaries who would, in turn, pass them on to the producer.
The latter assumption is reasonably applicable to the present
economic system, but the former is improbable. The official rate,
being set in the capital, is not likely to change because of road
improvements in the southern portion of the province. Furthermore,
the failure to account for costs gives the truckers no knowledge of
what should be passed on to producers and almost any apparent

gain can be rationalised with the high rate of inflation because costs are also rising.

The best cost data for vehicle operation and maintenance are those from *Transportes Argentinos, Plano de Largo Alcance,* which were taken from the American Association of State Highway Officials' pamphlet and modified to Argentine conditions.† It is assumed that after a road improvement the highway will remain in its improved state and that traffic will continue at the same rate. Actually, both of these assumptions are contrary to the recent experiences in Puan, as highways have deteriorated and traffic has increased with time. The money costs are no longer applicable so they have been converted to indices.

Vehicle	Paved	Road gravel	Dirt
Truck (10–12 ton)	54·9	80·1	100
Truck (25 ton)	53·8	80·1	100

Since there is no appreciable difference in relative costs of operation for trucks of unequal tonnage on any one type of road, the costs will be considered to be in a 100:80:55 relationship for all trucks. That is, the costs of driving on paved roads and gravel roads are 55 and 80 per cent respectively of the costs of driving on dirt roads.

With these cost data the production within the radius of influence of Darragueira was examined. Darragueira was chosen because it is the marketing centre in the *partido* most distant from Bahia Blanca and because it is necessary to travel more provincial roads before arriving at Route 35. It is assumed that Route 35 is paved; that all provincial roads are gravelled; that products are hauled by truck; that traffic will be diverted to the new pavement; and that movement from farms to town is on unimproved local roads. With these assumptions the greatest amount of cost decrease should be demonstrated so that the theoretical model will indicate the largest gains which could accrue to truck drivers and be passed on to producers. (This excludes losses arising from delays caused by rainfall or 'tie-ups' at the port, so that on some occasion losses would be greater.) The above indices are multiplied by the number of kilometres of a section of highway of a given quality to arrive at the results in Table IV. It is demonstrated that with a national and provincial road

† *Transportes Argentinos, Plano de Largo Alcance,* Ministerio de Obras y Servicios Publicos, Buenos Aires 1962, App. I, 41.

construction programme (a more ambitious proposal than can be expected), the operating costs for vehicles could be reduced 18·9 per cent. Since traffic diversion requires more kilometres to be travelled than the present route direct to Bahia Blanca, the savings would be slightly less. Even if 20 per cent is accepted as the upper limit for savings for truck drivers and the entire amount is passed on to producers, this is applied only to the 8 per cent of the product sales prices or the 10 per cent of production and marketing costs which transport comprises. With the optimistic assumptions that have been made the decrease in total costs would not reach 2 per cent. It is this 2 per cent which can be viewed as an increase in the quantity of production marketed since it is no longer necessary to pay this amount (in kind) for transportation. A problem has been avoided by working with the marketing centre which is most distant, Darragueira, and showing that the cost reductions are relatively insignificant. The areas nearer to Bahia Blanca and Route 35 would show smaller cost reductions. Thus, while precise statistics are not available for the *partido*, the upper limit of cost reductions has been calculated and, because it is so low, the more detailed figures for intermediate areas are not necessary.

The estimated maximum cost reduction arising from highway improvements is not a large percentage of total costs and will not be in the future. The possible reduction in costs of production ranges

Table IV. Reduction in Transportation Costs between Darragueira and Bahia Blanca as a Result of Improvement of Provincial Roads and Route 35

Section of Road	Cost Index × Kilometres before Improvement	Cost Index × Kilometres after Improvement	Percentage Reduction
Farm to Town (15 kilometres)	1,500	1,500	0
Darragueira–Bordenave (15 kilometres)	1,500	1,200	20
Bordenave–La Pampa Border (44 kilometres)	4,400	3,520	20
La Pampa Border–San German (44 kilometres)	4,400	2,420	45
San German–Bordeau (76 kilometres)	4,180	4,180	0
Bordeau–Bahia Blanca (8 kilometres)	800	800	0
Total	16,780	13,620	18·9

Sources: Data from *Transportes Argentinos, Plano de Largo Alcance* applied to the roads between Darragueira and Bahia Blanca.

from 1 to 2 per cent, if the production climate remains unchanged after road construction. Thus, if road improvements are to increase agricultural production substantially, it must be shown that the time and cost reduction in sending products to market can be used advantageously, or that the improvements will promote a higher output in some other way.

INTERVIEWS IN PUAN

The use of these savings in time and resources and other possibilities of increasing production which transport makes available were examined on the basis of interviews with producers in Puan.

An important means by which highway improvements can lead to increases in agricultural production is by facilitating the sale of products which are now marketed. In some zones of the country poor transportation is responsible for higher costs and physical losses and, consequently, reduced production. In Puan this is most easily caused by inclement weather and/or poor roads which make the hauling of goods difficult and expensive. The farmers were asked, 'Could you sell more of what you produce with all-weather roads?' so that the strength of this factor could be determined. The responses did not indicate that roads act as a significant limitation on the quantity of products sold. Only seven of the sixty-eight respondents believed that road improvements could lead to changes in sales. It is interesting that no one suggested that more cattle could be sold, but only that higher prices would be received for the same production. Also, no one expected changes in either the quantity of crops sold or their prices after roads were improved.

Although it appears that roads do not strongly inhibit the sales of the goods which are now produced, road improvements can still have an appreciable effect on the level of production. This can be conveyed either by new opportunities or by releasing resources which can be used more productively. The producers were asked, 'Could you produce more with all-weather roads?' to examine the strength of the stimulus of road construction. The question referred to both increases in output of the goods presently produced and the introduction of new products. It was expected that answers to this question would be overly optimistic because of the general faith in roads as a crucial element in marketing. However, only fourteen of sixty-eight replied affirmatively. This is important evidence that producers think that they are now producing all that they are capable of marketing and, in combination with the answers to the previous question, is evidence that all that is produced is sold.

The failure of more farmers to anticipate greater production is probably attributable to the geographic limitations of the area within which a farmer has little land on which to introduce new products and usually must be content with shifting between cattle and crop production. In all, eight farmers indicated that better roads would lead to more crops since it would be easier to transport the harvest. No one mentioned producing more cattle so roads evidently do not limit livestock production, although, as was seen previously, they do limit the markets to which animals can be sent. Even if there is a shift from cattle to crops, an increase in production is not certain since cattle, perhaps, were originally produced relatively less inefficiently. If a producer changes from less than optimal cattle ranching to less than optimal wheat farming (methods in both areas of production are typically less than ideal), it is difficult to estimate the net gain or loss in production.

The widespread failure to affirm that there are possibilities to increase production and sales, even though some expect small cost reductions, could substantiate the contention that the producers do not believe that they lack modern methods. Alternatively, it might imply that the producers consider any improvement for example, heavy machinery, too expensive to be offset with gain from cost reduction. Either of these beliefs makes agricultural extension work difficult since a producer cannot be convinced that a change in his techniques is desirable if he believes that he already uses the best available and that he could be aided only by the application of costly innovations.

Only sixteen of sixty-eight producers (less than one-fourth) affirmed that they would change production methods if roads were improved. Perhaps the failure of more producers to anticipate changes is because road improvements in the area would not have a strong effect on production or because it is thought that there are no available methods which are unutilised. There was a recurrent opinion in the interviews that road improvements would not benefit the interviewee personally, but that sizeable benefits would accrue to others and that the interviewee used the most modern methods, but that other producers did not.

The changes in methods in ten of the cases were not in farming techniques, but rather in shipping more by truck to Bahia Blanca. The remaining six believed that more modern techniques and more machines would follow from road improvements. The failure of many to foresee changes stems from the fact that a number of farmers do not need improved roads for marketing their production. Truck drivers are able to reach the farms and the products can be sold in the towns a short distance away. Poor roads are an inconvenience, but usually not a costly one.

It is likely that some changes in production methods will require a longer period of time to be initiated. However, only nineteen of sixty-eight felt that more changes in methods would occur over time as a result of road construction. Furthermore, not all of the respondents believed that the results would be favourable. Some producers on smaller farms thought that transportation improvements would enable more people to live in larger towns or cities and lead to the vanishing of the small towns and the close source of supplies which they provide. The general belief was that social effect of road building would be far-reaching and that these would somehow lead to the application of improved techniques. Neither the specific techniques nor the mechanism by which the application would be conveyed were suggested. Those producers with farms closest to Bahia Blanca anticipated no changes in techniques even after a lapse of time so that benefits, in whatever manner they are transferred, perhaps will be greater in the more rural zones.

It is not expected that the responses relevant to paving Route 35 would show predicted increases in production since questions concerning improvements in highways in general did not result in expectations of production changes. The answers of the producers, however, yielded some interesting relationships. The producers were asked, 'What would you do differently if the 44 kilometres of Route 35 between San German and Nueva Roma was paved?' With respect to the section between San German and La Pampa forty of sixty-eight persons expected no production changes. These forty were, with a few exceptions, situated in the northern or southern extremities of the *partido*, or had farms very close to Route 35. In the north and south there were no plans to take advantage of the new pavement, often because of the distance involved, while those near the road felt that the road already allowed adequate transport. The farmers who had highest hopes for better conditions were those living from five to fifty kilometres from the highway who believed that diverted traffic and more conveniences would follow its paving. Only four of these producers, however, foresaw production increases.

The expectations accompanying the prospective pavement can be checked against responses concerning the paved section between San German and Nueva Roma. The section Nueva Roma and San German is the middle of the three between Bahia Blanca and La Pampa. The paving was completed in 1964. In this case, also, there were negligible effects in the northern and southern extremes. Not many of the producers near the paved segment were interviewed since most of the section is not in Puan, but those who were within about twenty kilometres of the section felt that they received benefits from the construction. Probably, as some indicated, those between twenty and fifty kilometres do not receive the benefits that they expect without simultaneous improvements in feeder roads. The number of producers who can reach the pavement depends on the quality of 'secondary roads', in any case, but generally it appears that better conditions would accrue mostly to those producers located between, perhaps, five and twenty-five kilometres from the new construction.

The general opinion seemed to be that roads are desirable and it is expected that they will have an appreciable effect on the economy, but that this effect will not be through increased agricultural production. With agriculture accounting for 75 per cent of the *partido's* output, and services to people in this sector comprising a large share of remaining product, it is difficult to see how this would otherwise occur. There are long-run improvements possible, but these cannot be expected to be great according to the earlier responses of the producers in reference to changes over time.

OTHER FACTORS IN PRODUCTION

It must be emphasised that while the opinions of producers are important, they are not the determining factors in production. Those farmers who did not foresee production increases might find their expectations change while those who aspired to produce or sell more with improved roads might find their actions hindered by an unanticipated factor.

It is assumed that all cost reductions are passed on to the agricultural sector, some production increases can be realised. New products or larger quantities of present products could result from new inputs or new combinations of inputs. The cost reductions though, are not of a magnitude which permit large increases in output. Even if they were, the attitudes of the producers indicate that few changes in output are anticipated. Without the belief that greater production can be obtained and the desire to produce more, cost reductions arising from an increase in the supply of transportation can be expected to have only minimal effects. Consequently, road improvements in Puan would not have a large effect on agricultural production and, as a result, most likely would not have a significant impact on the economy.

These results are applicable, with slight modifications, to the Bahia Blanca–Santa Rosa region. The heterogeneous physiographic features which are present in Puan make it very representative of diverse conditions in other parts of the region. Its disparate road qualities and soil types and varying quantities of rainfall permit the study of a wide variety of relationships between agricultural production and transportation. The producers are affected similarly in the areas closer to Bahia Blanca than Puan and between Puan and Santa Rosa. In livestock production there is no apparent reason why benefits should be greater elsewhere than in Puan and wheat producers in the *partido* probably fare better than most. Generally, there are few differences between farmers in Puan located twenty or thirty kilometres from Route 35 and similarly situated farmers in other *partidos*. In both cases the absence of a feeder road system reduces the benefits which could be received by improving the main highways. Furthermore, even the presence of better feeder roads would not significantly alter the situation since there are no evidences of 'arrested' desires to introduce new products or to change methods or burdensome problems in marketing production. Without more marketing difficulties and a relatively more important role of transportation, road improvements cannot precipitate large changes.

CONCLUSIONS

For Puan and the Bahia Blanca–Santa Rosa region it was concluded that highway improvements would not lead to a substantial increase in agricultural production. According to the technicians at the national agricultural experimental agency (INTA) there are substantial untapped production possibilities, but there is no indication that road improvements will be a key factor in their realisation.

George Wilson's comments are particularly cogent in this context: 'Transport is no more an initiator of growth than any other form of investment or deliberate policy. Sometimes it can be strategic, but so can any other type of investment' (Wilson, 1965). It is useful to review the reasons why transportation failed to be strategic.

It is important initially to distinguish between *new* road construction and road *improvement* undertakings. At the outset and *a priori* it appears that road improvements are unlikely to produce large economic benefits as compared to the potentialities of new construction. Nevertheless, opportunities for large production increases exist and presumably government planners conclude that road improvements will assist in their realisation since roads are built for this purpose.

Road improvements would be expected to have fewer effects in the Bahia Blanca–Santa Rosa region than in other areas. The region is equipped with railroad facilities and a system of dirt roads which provide access to both small, 'neighbourhood' towns and larger marketing centres. The main roads to be improved and the railroads are parallel and serve the same towns. Consequently, the improvement of roads furnishes few new transport alternatives and provides additional competition for an already moribund carrier. Only if road construction substantially lowered costs, released resources, or broadened alternatives would significant results be expected. Perhaps an 'opening-up' situation is necessary to satisfy these requirements. Only in the southern portion of the region does an 'opening-up' possibility exist and there the physiographic features set a restrictive upper limit on development, suggesting there is little there to 'open up'.

The physiographic characteristics are the constraints within which road improvements can be effective. In the Bahia Blanca–Santa Rosa region there are few evident opportunities to introduce new products so that, at best, the paving of roads could help expand production of present products on which roads now set no particular limitations. In the southern portion of the region, production could probably be expanded but only with higher unit costs due to expenses of preparing the land for production, uncertain climate, and limited soil capacity.

Within the physical environment, road improvements can exert an influence on agriculture primarily through the producers' actions. There are now opportunities available which are not used and their realisation is not limited by a lack of roads. Consequently, better roads probably would not alter the situation greatly. Perhaps the sizes of the land holdings or the systems of tenancy limit the reactions. Alternatively, the pace and level of living might be such that the psychic costs of changing them outweigh any prospective financial benefits. Whatever the reasons, there is a noticeable reluctance to change. Farmers have produced the same product for a number of years and are accustomed to 'time-tested' production and marketing techniques. Without stronger aspirations to change, expectations of effects of highway improvements cannot be optimistic.

Even if there existed a conducive physical environment and plans to take advantage of opportunities when they arose, the effects of road improvements on producers could be limited by marketing facilities. Without cooperating factors at the ends of a road, its paving will not produce favourable results. In this region production can always be sold, but the delays and added expenditures make it less profitable than would be possible. Road improvements must be accompanied by expanded livestock and grain marketing facilities if they are to achieve maximum effectiveness.

Investments must be viewed in relation to the complex of factors in which they operate. Perhaps this is more important in the case of roads since the physical presence of a road does little or nothing if vehicles are not available to use it. Furthermore, without feeder roads to connect a main highway to internal areas, the highway's effectiveness is impaired. In the Bahia Blanca–Santa Rosa region there has been no problem in expanding the trucking industry to carry the produce where roads exist. However, there has been a deficiency of well-maintained feeder roads. The responses to questionnaires reflected the belief that without a system of roads significant effects of paving Route 35 had not previously been felt. Also, effects were not expected to arise from future highway construction projects if feeder roads were not improved as well. A coordination of highway improvements with provincial roads and with other factors is necessary for maximum benefits.

The Bahia Blanca–Santa Rosa region is not representative of a situation typically associated with agricultural economies. There are no overbearing problems with capital invested outside the agricultural sector, impossible obstacles to overcome in production or marketing, or an appreciable absence of cooperating factors. Limitations on production arise, to an extent, from all of the above factors. None, however, appears to be so strong as the psychic costs which outweigh the benefits from the adoption of modern techniques. It is not likely that road improvements will assist the solution of this later problem.

Although there are limits to the benefits to be received, road improvements will provide some savings. Theoretically, the producers should be the recipients of gains from reductions in vehicle operations and maintenance expenditures and in transportation costs after they are passed on by trucking enterprises and 'middlemen'. The truck tariff structure and marketing methods do not insure that the theoretical expectations will materialise. The prospects of rate reductions following road improvements are not favourable, since the wheat truckers officially operate on the basis of rates set in the provincial capital and the cattle truckers on the basis of what the trucking enterprises decide. Cost savings need not be passed on to the producers with this system, but they would be 'enjoyed' by the society in the aggregate.

The small positive effects suggest that prospective road improvements should be examined closely and that postconstruction evaluations would be helpful. Perhaps more postpaving evaluations would lead to more prepaving studies by elaborating variables or showing unanticipated results or unjustified expectations.

The approach used here does not permit a decision on whether construction should take place since the economic and social benefits and costs arising outside of the agricultural sector and aspects of agriculture other than production are not considered. In many regions of Argentina, however, agriculture and the economy are nearly synonymous so that if road improvements do not lead to more agricultural output, they can only be justified in a more nebulous manner as social or consumption goods. If the economy cannot support expenditures on consumption goods, a more appropriate allocation of funds can be chosen. When the emphasis is on the highway rather than the goal, other pertinent investments are often ignored. If an increase in agricultural production is the desired goal and highway improvements are viewed as one of the ways to achieve it, other possibilities are not obscured. When road evaluations are made on a regional basis with particular goals in mind, perhaps the impression that roads are panaceas for many economic problems would be dispelled and realistic alternatives made evident.

REFERENCE

WILSON, G. (1965). *The Impact of Highway Investment on Development*, Washington.

7 The Importance of Passenger Transport in Nigeria

ALAN HAY†

MANY observers have noted the importance of passenger transport in the developing countries of West Africa. The qualitative impressions of such observers have been confirmed by the more careful students of the transport situation (Hawkins, 1958; Walker, 1959; Hay, 1968). In addition, passenger transport has been seen as important in studies of migration and urbanisation (Niddrie, 1954; Rouch, 1954; Prothero, 1958; Kuper, 1965). Despite such acknowledgement of the importance of passenger transport, there is no general description or explanation of the phenomena involved. This enquiry attempts to repair this omission with a description of the major patterns observed in Nigeria, followed by a general explanation in terms of the economy and society. This paper presents material collected in the period 1963–5 when the author was holder of a Commonwealth Exchange Scholarship at the University of Ibadan, and an Associate Research Fellow of the Nigerian Institute of Social and Economic Research. The author is grateful for the encouragement of Professors K. M. Barbour and A. L. Mabogunje, Dr R. H. T. Smith and Mr Michael Chisholm, and for the lively co-operation of those undergraduates who took part in the study. The map was drawn by Mr D. Orme at the University of Leicester. A debt of gratitude is due to members of the West African Studies centre at the University of Birmingham who offered their comments on an earlier draft of this paper. All errors of commission and omission remain the author's responsibility. This general explanation is then considered in the light of a passenger survey conducted at the Ogunpa motor park in Ibadan in 1965. The concluding section suggests consequences for both policy making and further research.

DESCRIPTION

The first impression of passenger transport in Nigeria suggests that it is an industry without well developed patterns; but the study of

† This is a revised version of Dr Hay's paper, published in 1969.

road transport survey data suggested that there are in reality four types of passenger transport, each with its characteristic patterns.

The phrase '*topping-up*' is used to denote the first of such types. In topping-up, the passenger is secondary to freight both in the space he occupies, and in the contribution he makes to operating revenue. The vehicle will provide wooden benches upon which passengers are seated, indeed the law insists that they should be given a minimum width of fourteen inches and a depth of at least ten inches.† Topping-up arises in three ways. Firstly, the passenger may have some commercial connection with the goods being carried; he is accepted as part of the total transport charge. Secondly, some passengers may be picked up at 'cut-rate' fares below those charged by the passenger transporter *sensu strictu*, but the service is correspondingly slower and more uncomfortable. Finally, in some areas where the demand for passenger transport is low, there may be no other available transport, and passengers accept it *faute de mieux*. Examples of these three aspects could be found between 1963 and 1965 (Table I); of

Table I. Passenger Traffic 'Topping-Up'

Survey	Traffic type	Percentage of passenger capacity occupied	Average number of passengers
Lagos (Iddo Park)	Over 5 tons	17	7·5
Ilorin	Through traffic		
	Northbound	24	10·4
	Southbound	15	6·9
Oturkpo	Through traffic		
	Northbound	21	n.a.
	Southbound	23	n.a.

Sources: Surveys by the author.

the first on the route between Onitsha and Jos, of the second between Ibadan and Kano, while the third could be observed on the routes between Keffi and Kaduna, and between Jos and Yola.

† Western Nigeria, *Road Traffic Law*, Ibadan, n.d., Cap. 113, p. 399, para. 45(1) subsections (*f*) and (*g*).

The benefit to the owner-operator of such passenger traffic may be direct or indirect, Directly it may be a form of marginal revenue, which exceeds both the marginal and opportunity costs of providing the passenger service. Indirectly it may represent a source of additional income to the driver; in which case the owner is able to pay a low wage to the driver, in the knowledge that perquisities are high.

Passenger transport by *mammy-wagon* is distinctive both in the type of vehicle used, and in the mixture of freight and passengers. The vehicle is usually in the two to four ton capacity class. The most popular makes are the BMC and Bedford, but on to the imported chassis is built a wooden covered body. This may be divided into two compartments—second class lying behind the driver's cab, and third class at the back, accessible by a rear door. This third-class compartment is furnished with wooden plank seats, which are placed in position after loads have been stowed. The first-class passenger will share the cab with the driver.

In many cases the loads are the property of the passengers, who are personally marketing their agricultural surplus, or are engaged in petty trade. In this case the fare is often negotiated for the passenger and load together. This allows great flexibility in charges which will vary according to the day-to-day fluctuations in transport demand. There is evidence that in this form of transport the passengers subsidise freight movement. If the passengers pay 1d./mile for their carriage alone, this is equivalent to a revenue of 9d./ton mile. In many cases the passenger fares drop to about 0·8d./mile, but the equivalent ton mile rate of 7·2d. still lies well above the predominant freight rates of between 4·5d. and 6·5d./ton mile. These calculations were made for a typical mammy-wagon of four tons capacity, and licensed to carry a maximum of thirty-six passengers (Table II). If the vehicle carries passengers and freight in equal proportions, the

Table II. Mammy-wagon Transport on the Ilorin to Kabba Road

	Number of vehicles	Average pass. cap.	% occupied	Average no. pass.
Outwards	84	31	43	14
Inwards	83	32	46	15

Source: Surveys by the author.

revenue pattern per mile is of approximately 14d. from the passengers and 10d. from the freight. This represents a cross-subsidy of freight amounting to at least 1d./ton mile. In addition, it should be remembered that passengers load and unload themselves (so reducing labour costs), while over-loading is of course the rule.

Minibus transport (Table III) is a more recent development in Nigeria. Several makes of vehicle are used, including the Volkswagen, Ford (Germany), the BMC J-class, Bedford, Fiat O-M, and Mercedes. The frequency and popularity of makes is closely related to

Table III. Loading Characteristics of
Minibuses at Ogunpa Park†, Ibadan

Year of survey	1963‡	1965§
Passengers carried		
1–3	(15)a	—
4–10	14	10
11–20	90	50
21–30	9	2

Source: Surveys by the author.

† This included vehicles leaving the park for fuel, etc.; such trips were excluded in 1965.
‡ Between 05.00 and 20.00 hours.
§ Between 07.00 and 19.00 hours.

the credit and hire-purchase policy of agents at a particular time and place. These minibus operations are marked by a very high capital utilisation. This is reflected in high mileages and high load factors. Mileages can exceed 500 miles in a day, but a more usual average use is 200 miles per day, or between 5,000 and 6,000 miles in a month. The higher figures will only be achieved when the demand is high, when demand falls the vehicle will be idle for longer periods awaiting passengers. Such waiting becomes necessary because minibuses will not depart until fully loaded. In order to reduce empty running and idle time, passengers are charged on an origin-to-destination basis. Intermediate passengers may be carried, but will be charged the full fare for the trip. With full loads and few intermediate halts, a long journey can be completed at an average speed of at least 40 m.p.h.

Finally there are the *inter-city taxi services*. These are best developed in the Eastern Region of the country, but also occur to a lesser extent in the West and Midwest, and on a very few routes in the North (Table IV). The emphasis is again on great speed, long hours, and high load factors. Where a high load factor cannot be guaranteed, the passengers are expected to charter the vehicle for the whole trip. The favourite vehicle is the Peugeot 404 which can be made to carry seven passengers. Mileages range up to 500 per day, and between 5,000 and 8,000 per month. At Asaba a survey stopped 129 taxis, of these 100 were performing a total of 9,218 miles outwards, all on round trips, so an average day's distance of 180 miles can be supposed. Fares are higher than for the other forms of passenger transport, lying between 1d. and 1·5d. per passenger mile. On some routes these taxis provide almost the only form of passenger transport, but on other routes they are in competition with one or more of the other forms.

The general explanation

An explanation of the high level of passenger transport demand in Nigeria is often given in terms of the marketing system. Three possible procedures are open to the farmer wishing to sell his produce. The maximum demand for passenger transport will occur when a farmer or his personal representative travels to an urban market with the produce. A lesser transport demand will occur when the producer is persuaded to sell to a middleman or forestaller near to the point of production. Passenger transport demand will be minimised in the third case: the use of a correspondent to whom the goods are consigned. This last approach pre-supposes a high degree of trust between the consignor and consignee, and in the reliability of the transporter. It has only become the dominant pattern in certain of the long-distance, inter-regional trades. The explanation of passenger transport demand in these terms alone is not acceptable in the present context for two reasons. Firstly, it allows the omission of important contributory factors, and secondly it would seem to be based on a fallacious distinction. Personal marketing and personal travelling are so closely associated that it cannot be argued that the one causes the other, on the contrary both are reflections of causes which lie deeper in the nature of the economy and society concerned. The four causes proposed here are intimately inter-related but are distinguished here for purposes of discussion.

Table IV. Loading Characteristics of 'Inter-city Taxi' Traffic

Survey	Asaba 1964†		Iddo 1965‡	
	licensed	carried	licensed	carried
less than 4	—	12	—	1
4	6	7	—	5
5	6	3	2	4
6	1	6	1	15
7	63	42	75	28
8	—	4	7	26
over 8	—	2	—	6

Source: Surveys by the author.

† 07.00–18.00 hours, 2 days.
‡ Continuous 2 days.

Most important is the paucity of long-distance communications. This is in a sense paradoxical, for as other authors have noted in Nigeria, indeed in all African countries, local intelligence is remarkably good. Over long distances however, poor communications result from poverty, illiteracy, and inefficiency. The first two reduce the number of people who can make use of the facilities available, the third makes communications slow, unreliable, and inaccurate. For example, in the General Strike of 1965, news of the settlement could not be transmitted reliably to the provincial strike leaders until emissaries from Lagos had travelled with the news. Attempts have been made to improve long-distance market intelligence with newspaper and radio bulletins of market prices; but the methods of data collection were so unsophisticated that no astute Nigerian accepted them.

The concept of the extended family is well established in the sociology of Nigeria. Evidence put forward by Marris, Izzett and others suggests that even the processes of migration and urbanisation do not destroy the loyalty to the extended family (Izzett, 1961; Marris, 1961). There is therefore a continued observance of the traditional obligations to attend naming ceremonies, funerals, and marriages. An excellent summary of such a pattern is given by Goddard in his study of rural–urban relations in Ọyọ (Goddard, 1965).

The concept of urban hierarchies also throws light on this problem. Grove and Huszar in Ghana, and Abiodun in Nigeria suggest

Table V. The Questionnaire Used in the
Passenger Survey at Ogunpa Park, Ibadan

Lorry-park station
Male or female
Type of vehicle
Time of day
Region of Birth
Province of Birth

Occupation:
  Agriculture
  Crafts
  Trade
  Clerical
  Administrative, professional or technical
  School or college
  Unemployed
  Other or unknown

How long have you been in Ibadan?
Normal place of residence?
Destination?
Fare?

Why are you travelling?
  To buy
  To sell
  Family business
  Education, medical, etc.
  Government business
  Seeking work
  Other or unknown

(1) As stated in the text, the interviewers were instructed to derive answers
    by direct and indirect questioning. The category 'other or unknown'
    thus includes three categories: (i) refusals, (ii) multiple motives with no
    dominant motive apparent, (iii) individuals whom the enumerators
    were unable to categorise due to confused answers.
(2) In many cases the motive was 'to return home': in such cases the inter-
    viewers were asked to elicit the purpose of the trip (non-Ibadan resi-
    dents only).

that such a hierarchy exists in West Africa, even though Abiodun
is unable to detect the presence of distinctive geographical patterns
(Grove and Huszar, 1965; Abiodun, 1967). But without such sophisticated formulations it is evident that the incomplete coverage of
medical, educational, and other services will involve long journeys
for those who need them. In some cases, this need is exacerbated by
the inefficiency of communication by postal and telegraphic services.

The most extreme example of this demand for travel are the specialist hospitals. Thus the Kano eye hospital draws patients from all parts of Nigeria and beyond. If the patient is accepted into hospital a secondary demand for the transport of visitors arises.

Linking all these and the question of personal marketing are the low opportunity costs of time. Time has no value *per se*, only the costs of the opportunities which are foregone. The opportunity costs of time in Nigeria are very low; this being associated with unemployment, concealed unemployment, and under-employment. This makes the fare paid the only cost for those who travel for social reasons or for services. The point can be illustrated from personal trading. An Ibadan trader may learn that a line of merchandise is cheaper in Lagos. The fares for his small boy (often a son, nephew, or other relation) to travel to Lagos is ten shillings return. If he can return with say one hundred items, saving only two pence on each item, it will yield him a net profit in excess of six shillings. If the time so spent has no notional value, the operation is profitable. P. T. Bauer reports a similar procedure in *West African Trade*, London, 1963.

If a general explanation in these terms is correct, it can be tested against data from the accurate observation and survey of passenger travel. Such a test can be conducted at two levels: macro- and micro-levels. On a macro-scale it would be reasonable to expect relationships of the total passenger flow between urban centres and their principal demographic, social, and economic characteristics. For example, a relation might be observed between passenger demand and the level of migration into a town in recent years. Such analysis has been attempted by Snyder, 1962.

Testing the general explanation

It was possible partially to investigate the proposed explanation at a micro-level: seeking these relationships through a pilot survey of the characteristics and motives of individual travellers at the Ogunpa Motor Park in Ibadan. Essential background to this survey can be found in Lloyd *et al.*, 1967. This park has the advantages (for such a study) of being enclosed, so that vehicle departures can be recorded. In addition the park is clearly divided into 'stations' or 'sides', recognised by convention. The disadvantages lie in the fact that in 1965 (the time of the survey) it was carrying only a small part of the traffic by taxi between Ibadan and Lagos. Further, much of Ibadan's mammy-wagon traffic does not use the park, but terminates at or near the great urban markets at Bere, Dugbe, Mokola, etc., this being especially true of the shorter-distance mammy-wagon traffic from destinations less than thirty miles from Ibadan.

The questionnaire which is reproduced was administered to travellers by a team of final-year students at the University of Ibadan. The categories of the questionnaire were defined and discussed before the survey, the interviewers were asked to derive results by direct and indirect questioning. All students spoke Yoruba and English; those interviewing north-bound traffic also spoke Hausa. Travellers were informed that the inquiries were academic rather than governmental in character. The problems of arranging such a survey in the hurly-burly of a lorry park made it difficult to stratify, randomise or otherwise ensure a representative sample: the only stratification was by allocation of the interviewers to motor park 'stations' (divisions recognised by convention and checked against the declared destination of travellers). In Table VI some of the shortcomings of

Table VI. Ogunpa Park Survey 1965. Travellers and Interviews

Sex	Departures		Interviews		Interviews as percentage departures	
	Male	Female	Male	Female	Male	Female
Station:						
Lagos	485	309	94	84	19	27
Iwo & Ife	415	327	104	82	25	25
Ilorin	183	81	44	48	24	60
Ijebu-Ode	207	143	79	62	38	43
Abeokuta	158	145	54	31	34	21
	1,448	1,005	375	307	26	30

the interviewers can be identified: including the marked predilection for the ladies on the Ilorin station. Having identified this source of error, it can be noted that no group fell below thirty interviews, and that only in one case was less than twenty per cent of a group interviewed. It is because of these recognised defects in the sampling and the large number of refusals to certain questions, that only the grosser variations in succeeding tables are accepted as significant.

The position of Ibadan as the administrative, commercial, and in many ways the cultural capital of the Western Region would lead us to expect an excess of travellers inwards (Lloyd *et al.*, 1967). This is confirmed by the residence of travellers as shown in Table VII: the exception is the Lagos station, on which a large number of Ibadan residents are recorded. The flow to Lagos will consist of three groups: Ibadan residents, Lagos residents, and people from other places travelling to Lagos via Ibadan. Thus the most important features of Ibadan's position in the urban hierarchy find an immediate expression in the patterns of travellers residence.

Table VII. Ogunpa Park Survey, Residence of Travellers

Residence	Number Ibadan	Elsewhere	Percentage Ibadan	Elsewhere
Men	74	267	22	78
Women	73	214	24	74
of whom, Lagos travellers only:				
Men	32	57	36	64
Women	32	49	40	60

The motives for travelling as recorded in Table VIII also confirm the general explanation. The close association of personal travelling with personal trading and family obligations is evident for both men and women. Closer examination of the trading motive reveals differences between the sexes. In all cases the family motive was dominant, but it was significantly greater for women than for men. It also appears that men are significantly more likely to be travelling to buy, women to sell. Fourth motive, in order of importance for both sexes, was the use of educational, medical, and other services. Among those who had some government business to discharge, men outnumbered women by four to one.

Consideration of the variation in motive according to station on the lorry park reveals a clear reflection of the urban hierarchy. There is a significant difference between destinations south of Ibadan (where buyers exceeded sellers), and destinations north of Ibadan (where the reverse obtained). It is not possible to advance any convincing reason for the variations between the stations in respect of the educational-medical motive. It can only be noted that with such

a small sample over a single day, the accidental coincidence with a
school holiday might interrupt an otherwise uniform pattern.

Table VIII. Ogunpa Park Survey, Motives for Travelling

Percentages:

Station	Abeokuta	Ijebu-Ode	Iwo-Ife	Ilorin	Lagos
Motive: Women					
To buy	7	13	9	7	18
To sell	13	25	25	23	8
Family	40	47	53	37	60
Education, etc.	17	6	3	0	10
Government	0	0	0	3	4
Seeking work	0	0	0	7	0
Other	20	9	7	23	1
Total	100	100	100	100	100
Number	30	57	79	30	80
Motive: Men					
To buy	18	24	12	18	14
To sell	6	4	10	29	14
Family	22	37	41	47	42
Education, etc.	8	9	19	0	3
Government	20	6	3	0	10
Seeking work	4	1	6	0	8
Other	22	18	9	6	9
Total	100	100	100	100	100
Number	49	79	88	34	87

The occupational pattern of the travellers is recorded. The most
dominant group by far are the female traders. This needs some
explanation; in Nigeria every active woman not otherwise employed
will seek to participate in trade. Although her turnover may be
limited to a few pounds per month she will nevertheless be termed a
trader. The most numerous group of men is in the administrative-
professional category as well as in trade. There were no significant
differences between stations in occupational pattern.

The simultaneous survey of vehicles leaving the park recorded the
destination, the number of men and women, and type of each vehicle
leaving the park. In addition to allowing a comparison with the
interview data it allowed further tabulations. The average length of

journey was revealed as about eighty miles; the modal distances being eighty to ninety, forty to fifty, and fifty to sixty miles. Few trips over 200 miles were recorded. In the same tabulations it was evident

Table IX. Occupation of Travellers, Ogunpa Survey

Occupation/Sex	Numbers		Percentages	
	Male	Female	Male	Female
Ag. and Fish	41	10	12	3
Crafts	27	11	8	4
Trade	81	168	24	57
Admin. prof.	90	31	26	11
School, etc.	29	20	11	7
Unemployed	9	21	3	7
Other	30	29	9	10
Clerical	24	4	7	1
	341	294	100	100

that there was no significant difference in the types of vehicles chosen. No taxis were recorded; about 28 per cent of passengers travelled by minibus. Most of the remainder travelled by mammy-wagon, only seventeen vehicles were licensed to carry more than forty passengers and they carried on average twenty-two passengers, indicating mammy-wagon rather than topping-up type of operation.

CONCLUSIONS

The Ogunpa data confirmed much of the *a priori* explanation. It is now necessary to examine two further topics, the scope for further investigation, and the implications of the argument so far.

In terms of research, there are many problems unanswered. The high scores for 'other motives' arouse one's curiosity; in addition, there would be scope for an investigation which acknowledged and evaluated the presence of mixed motives. It may be necessary to distinguish between the reason for the trip and the occasion of the trip. Research on seasonal fluctuations would also be of great value. Qualitative evidence suggests that at the height of the export crop harvests the number of rural residents travelling will fall, but the amount of trading activity will increase. Unfortunately it is impossible to investigate these until political conditions in Nigeria are more settled. There is also the broader field, the correlation of inter-urban flows with urban characteristics: the macro-scale which was mentioned above. In the meantime, the lorry park and the travel-patterns of families provide great scope for further field work. As more data is collected the findings of this paper will inevitably be corrected and extended.

Finally, it is possible to note the consequences of this study in practical terms. There is now firm evidence for the important role which passenger transport plays in social and commercial life. Without it intelligence of all kinds would move more slowly, affecting the diffusion of new ideas so vital to the development process. In addition, it provides a cheap transport medium for the distribution of manufactured goods. Not only is the system cheap and (in a sense) efficient, but the element of cross subsidy already discussed is important. The effect of such reduced transport costs in stimulating agricultural production or industrial sales is problematical: it cannot for that reason be ignored. In another sphere as well road transport is important, it offers an opening to Nigerian entrepreneurs, a tradition which can be traced back through the writings of McPhee and others in the 1920s (McPhee, 1926; Ijewere, 1958; Hay, 1967). This has become more important as passenger transport allows participation with lower capital outlay, and lower overheads.

The 'rationalisation' of the passenger transport industry has been attempted in Ghana, and similar policies have been recommended for Nigeria (Stanford Research Institute, 1961). Such policies cannot be undertaken lightly, for they will affect the working of the transport industry as a whole, the economy and society of Nigeria.

REFERENCES

ABIODUN, J. O. (1967). Urban hierarchy in a developing country, *Economic Geography*, **43**, 347–67.

GODDARD, S. (1965). Town-farm relationships in Yorubaland: a case study from Ọyọ, *Africa*, **35**, 21–9.

GROVE, D. J. and HUSZAR, L. (1965). *The Towns of Ghana*, Accra.

HAWKINS, E. K. (1958). *Road Transport in Nigeria*, London, pp. 30–2.

HAY, A. M. (1967) The development of road transport in Nigeria 1900–1940, *The Journal of Transport History*.

—— (1968). *The Geography of Nigerian Road Transport*, Ph.D. thesis, University of Cambridge.

IJEWERE, G. O. (1958). *Rail and Road in Nigeria*, B.Litt. thesis, Oxford.

IZZETT, A. (1961). Family life among the Yoruba, in Lagos, Nigeria, (ed.), A. Southall, *Social Change in Modern Africa*, London.

KUPER, H. (ed.), (1965). *Urbanization and Migration in West Africa*, Los Angeles.

LLOYD, P. C., MABOGUNJE, A. L. and AWE, B. (eds.) (1967). *The City of Ibadan*, Cambridge.

MARRIS, P. (1961). *Family and Social Change in an African City*, London.

McPHEE, A. (1926). *The Economic Revolution in British West Africa*, London.

NIDDRIE, D. (1954). The road to work; a survey of the influence of transport on migrant labour in Central Africa, *Rhodes-Livingstone*, 15, 31–42.

PROTHERO, R. M. (1958). *Migrant Labour from Sokoto Province, Northern Nigeria*, Kaduna.

ROUCH, J. (1954). *Migration in the Gold Coast*, Accra.

SNYDER, D. E. (1962). Commercial passenger linkages and the metropolitan nodality of Montevideo, *Economic Geography*, 38, 95–112.

STANFORD RESEARCH INSTITUTE (1961). *The Economic Co-ordination of Transport Development in Nigeria*, Menlo Park, California, pp. 187–9.

WALKER, G. (1959). *Traffic and Transport in Nigeria*, London.

8 Recent Railway Construction in Tropical Africa

A. M. O'CONNOR†

IT is generally agreed that transport facilities have an important part to play in the process of economic development, in tropical Africa as elsewhere. However, there is much less agreement on the exact nature of their role, and especially on the likely impact of new investment in this field. It is sometimes suggested that improvements in transport facilities will themselves stimulate new forms of economic activity, but there is a contrary view that investment of this type is not directly productive and should be undertaken only when a clear need for it arises. The development plans which are now being implemented in nearly all African countries differ very greatly in the degree of priority given to transport, and also in the priorities accorded to the various alternative media. In this paper attention is focused on railway construction, and its contribution to African economic development today.

THE ROLE OF RAILWAY BUILDING IN THE PAST

The situation today undoubtedly differs from that 70 years ago, when the vital importance of railway building for economic development was quite obvious. While river transport played an important role in opening-up trade routes in some parts of tropical Africa soon after the partition of the region among the colonial powers, it was the building of railways which did most to make possible the movements of people, and particularly goods on a vastly greater scale. Railway construction made an immeasurable contribution to the revolutionary changes which swept across Africa as it became integrated into world politics and the world economy. The railways made practicable, for the first time in most countries, the export of a wide range of produce and the import of many goods from Europe. They also played a vital part in the establishment of colonial administration, and indirectly in the subsequent process of social change. It was

† This is a revised version of Dr O'Connor's paper, published in 1969. For the related Select Bibliography, see pp. 149–50.

A. M. O'Connor

this situation which led Lugard to say in 1922, 'The development of Africa can be summed-up in one word—transport'.

Table I. Tropical Africa: Railway Construction 1956–71

Country	Line	Length in km	Date of completion
Angola	Sa da Bandeira-Serpa Pinto	508	1961
	Dongo-Cassinga	80	1968
	Luanda-Maquela	c. 500	in progress
Cameroon	Yaoundé-Bélabo	296	1969
	Bélabo-Ngaoundéré	327	in progress
	Mbanga-Kumba	31	1969
Congo (B)	Dolisie-Mbinda	283	1962
Ghana	Achiasi-Kotoku	80	1956
Guinea	Conakry-Fria	143	1958
	Kamsat-Sangarédi	136	in progress
Liberia	Bomi Hills-Mano River	83	1961
	Buchanan-Mount Nimba	267	1963
	Monrovia-Bong	77	1964
Malawi	Mpimbe-Mozambique border	112	1970
Mauritania	Nouadhibou-Tazadit	650	1963
Mozambique	Nova Freixo-Malawi border	c. 90	1970
	Nova Freixo-Vila Cabral	263	1969
	Inhaminga-Marromeu	87	1969
	Caldas Xavier-Cabora Bassa	226	in progress
Nigeria	Kuru-Maiduguri	635	1964
Rhodesia	Mbizi-Nandi	90	1965
Sudan	Er Rahad-Nyala	c. 700	1959
	Babanusa-Wau	c. 450	1962
Tanzania	Mnyusi-Ruvu	190	1965
	Kilosa-Kidatu	109	1965
	Dar es Salaam-Zambia border	c. 950	in progress
Uganda	Kampala-Kasese	333	1956
	Jinja-Bukonte	68	1961
	Soroti-Pakwach	343	1964
Zaire	Kamina-Kabalo	450	1956
Zambia	Kapiri Mposhi-Tanzania border	c. 900	in progress

For the first 25 years of this century, the railways were un-challenged as major means of long-distance transport for both people and goods; and for a further 25 years the challenge of road transport had little impact on the role of the railways in most parts of tropical Africa. Since 1950 most of the railways have begun to suffer from road competition, and in certain countries this has now become very severe, even to the extent of contributing to rail closures as in Sierra

Leone in the early 1970s; but over the region as a whole they are still the chief mode of transport for goods in terms of ton-mileage, and locally they are still important for passenger travel. Indeed, in contrast to the situation in many other parts of the world, rail traffic is still expanding rapidly in most tropical African countries, even though generally less rapidly than road, and now also air, traffic.

RECENT RAILWAY CONSTRUCTION IN TROPICAL AFRICA

Furthermore, new railway building is still taking place. The main period of construction lasted from about 1885 to about 1935, after which little further development took place for 20 years; but since the early 1950s there has been renewed activity, and almost every year has witnessed the opening of some new railway. The various individual lines are listed in the accompanying table. They differ one from another in their character, and especially in the purpose for which they were built, but some regularities in the pattern of new construction can be seen. With reference to a longer time span and to all forms of transport, these regularities have already been suggested by Taaffe and others.

Several of the new railways follow the traditional pattern of a simple line leading inland from the coast, and in most cases these are intended to assist in mineral exploitation. The longest of these is that in Mauritania linking the iron deposits around F'Derik and Tazadit with the port of Nouadhibou. All the lines in Liberia are similar in nature and purpose. A larger number of new railways represent simple extensions to, or branches from, existing primary lines. Some of these were also built to tap mineral deposits, as in Uganda, Congo (Brazzaville)–Gabon and Angola. Others were designed to assist agricultural development, or conceived in even more general terms as agents for 'opening up' hitherto little-developed regions. These include two of the longest extensions to existing systems, those in north-east Nigeria and in southern and western Sudan.

These new primary lines and simple extensions account for most of the recent railway building in tropical Africa. In general, they represent a continuation of what has often been termed the 'colonial' pattern of links between the interior and coastal ports. There are still few of the lateral lines which produce true networks, such as are found in more highly developed regions, including even South Africa

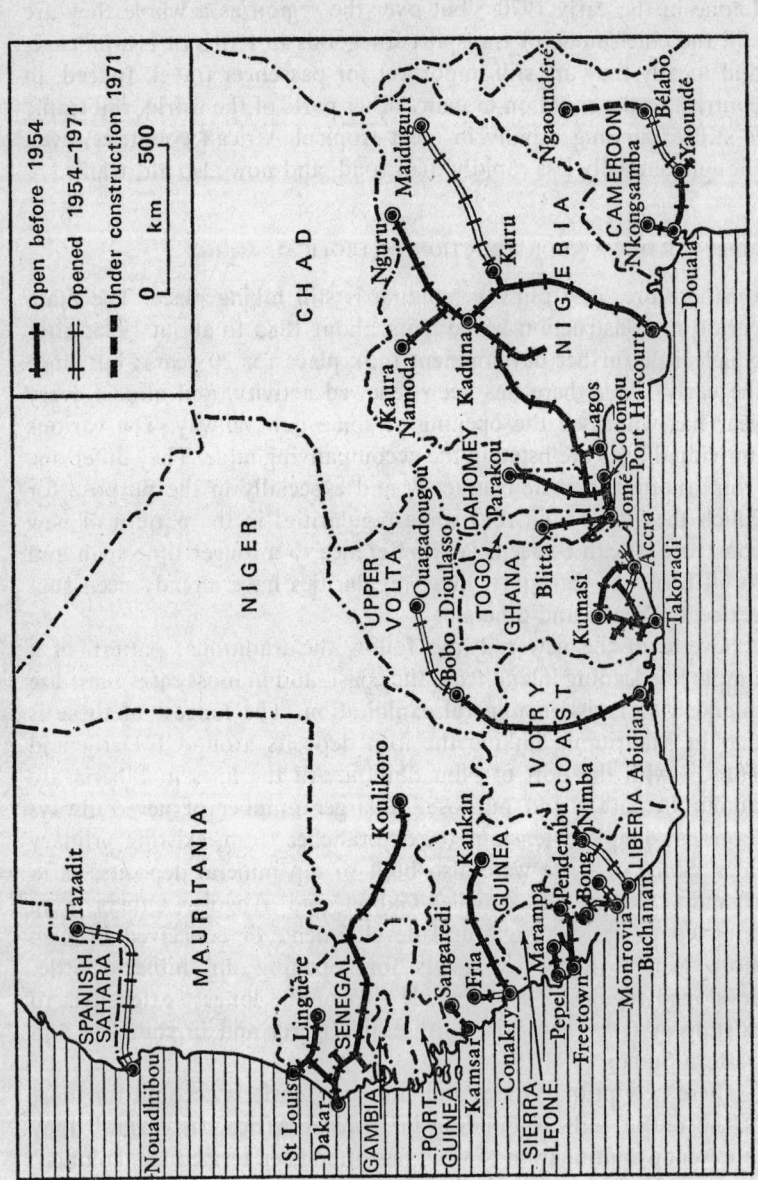

Fig. 8.1 Western Africa: railways

and the Maghreb. Among the few which do exist, two were opened as early as 1956, one in Ghana which provided a direct link between Accra and Takoradi (Fig. 8.1), and one in Zaire, which connected Katanga with the eastern African rail/lake system (Fig. 8.2). The only important lateral line built more recently is that linking the main railway across central Tanzania with the Uganda–Kenya–northern

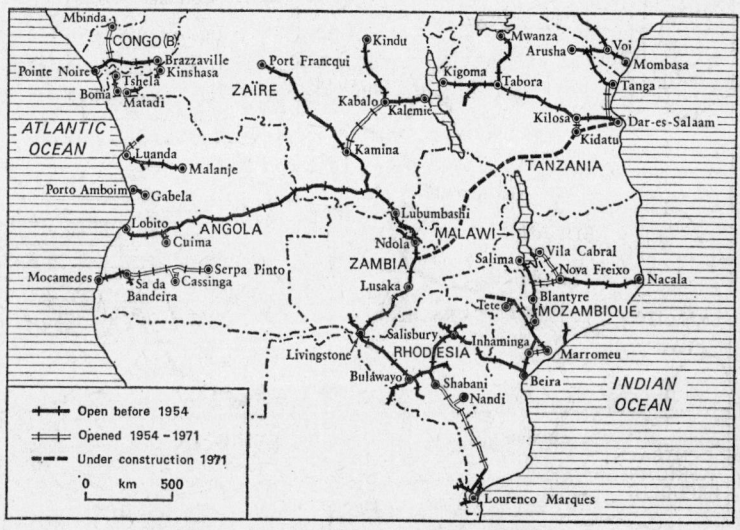

Fig. 8.2   South-Central Africa: railways

Tanzania system (Fig. 8.3), and this was justified more by arguments of convenience in East African railways operations than by traffic potential.

Thus, although the attainment of independence has brought much greater interest in the improvement of internal and intra-African communications, little has yet been achieved in this field as far as rail transport is concerned. More has been done to improve the road connections between neighbouring countries as well as internal road networks, and to reorientate air routes, which, in contrast to rail and even road routes, can be achieved at virtually no capital cost. It is relevant in this connection that intra-African contacts at present involve more movements of people than of goods.

Only two major new railways were actually under construction in 1971 in independent tropical African countries, although several more routes were being investigated, and even surveyed. The railway under construction in Cameroon represents a simple extension of an older line, designed to assist the economic development of one of the poorer parts of the country. The other substantial new rail-building project linking Zambia with Dar es Salaam in Tanzania falls rather

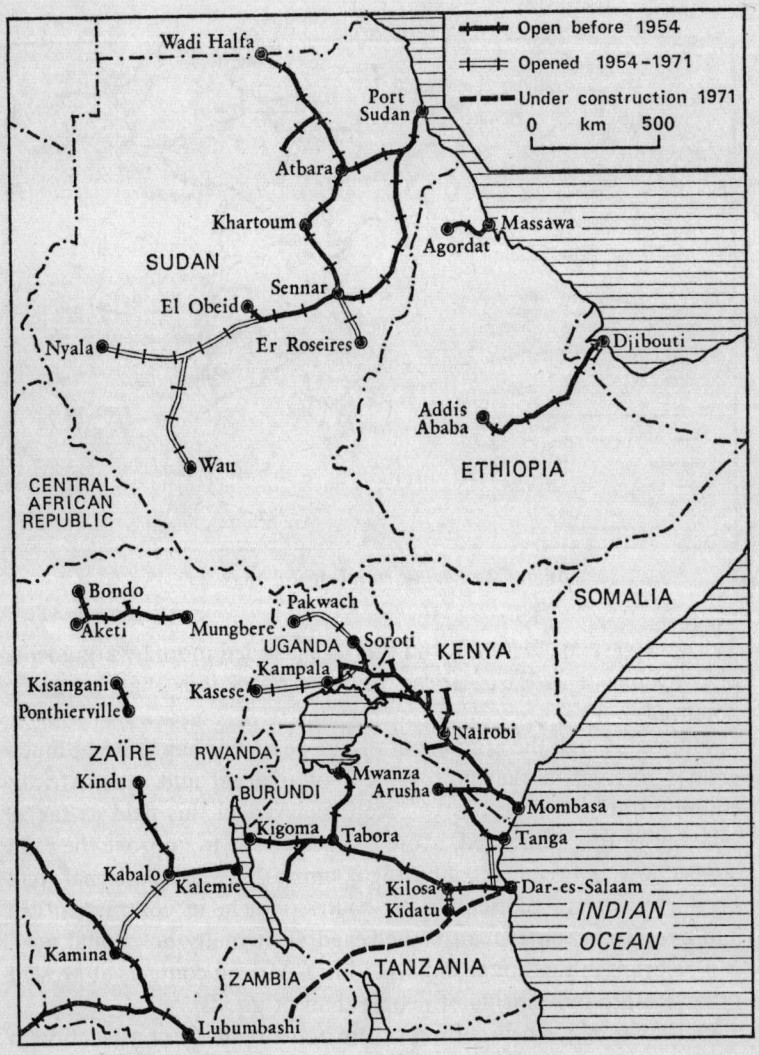

Fig. 8.3 Eastern Africa: railways

in between the categories of primary coast-interior lines and lateral connections. It is intended primarily to provide the Copperbelt with a greater choice of outlets to the sea, but it is hoped that it will assist trade between the two countries, and also 'open up' relatively isolated areas within each.

Some of the railways tentatively planned for the 1970s are clearly intended mainly to provide connections between neighbouring countries, such as the possible link-up of the Sudan, Cameroon and Nigeria systems. Others would form lateral links within countries, such as north–south lines in Angola. However, the lines most likely to be built are still mainly primary lines leading inland from the coast, and simple feeders to such lines. A clear example is the projected line from the Belinga iron deposits of Gabon to the new port of Owendo. One such line is in fact now being built to serve the Boké bauxite development in Guinea. If the recent boom in the economy of Ivory Coast continues, the plans for new railway construction there, designed mainly to assist agricultural and forestry development may also come to fruition, probably in the form of a primary route from the new port of San Pedro.

THE IMPACT OF NEW RAILWAY CONSTRUCTION

The main purpose of recent railway building in tropical Africa has clearly been to assist in the economic development of areas which previously suffered from inadequate communications, and most of the projects for the near future have the same objective. There is, therefore, surely a case for investigation to be made on the impact of the railways that have been built, for the results of this might be very relevant for the appraisal of future schemes. Ideally perhaps, this paper should provide an assessment of the role that each of the major new railways has played in the recent development of the area which it serves, although in some cases of course insufficient time has elapsed for any judgement to be made. It is, in fact, only possible to pose questions of this nature rather than to answer them: for very little evidence is yet available on the effects even of such major extensions as those in Sudan and Nigeria. The latter has been discussed by Barbour, but he too has formulated, rather than attempted to answer, the question.

A study of the western extension of the railway in Uganda,

undertaken in 1960–3, indicated that while it was successfully fulfilling its primary purpose of permitting the exploitation of the Kilembe copper deposit, its impact on other sectors of the economy of western Uganda was much smaller than had been anticipated. The situation does not appear to have changed greatly since 1963, while the north-ward extension of the railway system, completed the following year, has also brought about few striking changes in the area through which it passes. It is handling sufficient freight to justify its construction, but this represents traffic diversion from an antiquated river and lake transport system rather than traffic creation.

In several respects, conditions in south-west Sudan were particularly favourable for railway construction, notably because of the easy nature of the terrain and the exceptionally poor road network. Undoubtedly the new railways will be handling all the traffic that is being generated in the areas which they serve, but there have been no reports of significant new developments there. One important factor in this case is the political strife between north and south, which probably encouraged the building of the railways to assist national integration, but which then tended to discourage investment in the south.

In Nigeria, too, political strife has complicated the picture, the railway services having been severely disrupted by the civil war. But certainly, there too, there are no reports of any dramatic immediate results of the opening of the Bauchi–Bornu extension. In some respect, the situation there was less favourable than in Sudan, especially since the area had already been 'opened up' to a considerable degree by road transport. Furthermore, in Nigeria, as in Uganda, the hopes for a great impact on cash crop production ignored the fact that the presence of the railway does little to increase the farmers' incentive to grow cash crops when marketing-board policies ensure that a standard price is paid for these throughout the country.

It is reported that the branch from the Pointe Noire–Brazzaville line to the Gabon border is not only carrying manganese but is also assisting timber exploitation within Congo (Brazzaville); but this is not yet happening on any significant scale in Liberia, where a condition of the LAMCO concession was that the Mount Nimba–Buchanan railway should be open for public traffic. Some of the other lines built to link mineral deposits with the coast are, of course, not available for public traffic, and so cannot possibly have any direct impact beyond that upon the mining sector.

CRITERIA FOR ASSESSMENT

Even a superficial examination of these and other new railways helps us to distinguish the criteria upon which the decision whether to proceed with a new project should be based, although only intensive study of them will make it possible to evaluate these. One criterion is of course the nature of the terrain to be crossed, since this greatly affects both construction and operating costs. It is this factor which casts doubt on plans to provide Rwanda and Burundi with rail links to the sea, and on some of the proposals for extensions in Ethiopia. Even more significant, perhaps, are those factors which influence the present and potential demand for rail transport, including the physical and human resources and the present economic structure of the area served. One such factor is the present situation and the future policy with regard to other forms of transport.

In relation to any new railway project, therefore, there are many questions to be asked. Are there goods moving in and out of the area which would move more cheaply if rail transport were available? Are there goods such as food crops which now go to waste but which could be sold elsewhere if transport costs were lower? Are there resources, such as timber or minerals, which are not being tapped because of the absence of rail facilities? If, as is more usual, there are also other bottlenecks hindering development, such as inadequate disease control or power supplies, is the government willing and able to overcome these also, so that the opportunities provided by a new railway may be exploited to the full? If road improvements are also to be undertaken will these be complementary to, or competitive with, the new railway? Whatever the benefits that might be brought by a railway, could they be brought at lower cost by some other form of investment?

Some of the same questions must be asked in relation to link lines in so far as they are expected to have an impact on the area which they cross. Thus the prospects for the developmental effects of the Tanzania–Zambia railway within north-east Zambia seem very poor. Within parts of southern Tanzania they seem rather better, but only if the Tanzania government is prepared to allocate yet more funds to the south rather than to declare that the south has had its full share of public funds because of the railway construction. In such cases the analysis is more complicated, however, since much of the traffic handled is likely to be diverted from other routes, and since some of the benefits are seen in terms of security rather than lower transport costs. Indeed, there may be new lines which are clearly not justified on economic grounds, but which are probably justified if political considerations are also taken into account.

CONCLUSIONS

It is now generally realised that building new railways in tropical Africa will not produce economic miracles, especially now that so much of the region has already been 'opened up' to a considerable degree by road transport. Yet wild claims about the likely effects of such investment are still sometimes made, for instance in parliamentary debates. Undoubtedly, there are various circumstances in which the extension of rail transport to areas now lacking it is likely to result in the development of new forms of economic activity. One of these is the existence of mineral deposits, for which there is an assured demand, and which can only be economically exploited if a low-cost means of transport is available. It is for the movement of bulky commodities over a relatively long distance between two fixed points that rail transport is especially well suited, and it is certainly the lines carrying minerals which have made the most obvious contribution to economic growth in Africa in recent years.

There are also instances where new railway construction has clearly assisted agricultural development, especially where this involved the centralised production of a bulky crop such as sugar. On the other hand, there are probably only a very few areas in which new railway building is likely to stimulate a rapid increase in cash crop production on small farms on a scale sufficient to ensure that the traffic covers even operating costs. Certainly the impact of several of Africa's new railways on the bulk of the farming population in the areas through which they pass has been disappointing, and there have been lines such as that which operated in southern Tanzania from 1954 to 1962 which have been a serious financial liability rather than an asset.

There may be cases where the extension of railways to formerly remote areas would assist the growth of industry and trade in these areas, but it should be remembered that it could also discourage such growth, by increasing the competitive strength of products from existing industrial centres elsewhere. It can be argued that in some circumstances the improvement of transport facilities is more likely to lead to concentration than to dispersal of economic activity. Possibly, the new railways that are most likely to assist industrial development are in fact not those which are providing new areas with rail facilities; but rather those linking large centres which each have their own existing rail links to the coast, or which, in such cases as Lagos, Accra and Abidjan, lie on the coast. How far this is true depends very largely on the willingness of African countries to overcome the problem of limited markets by effective economic integration which would give industries free access to markets in neighbouring countries.

Undoubtedly, in the next decade more attention should be given to the possibilities of new, lateral rail links within tropical Africa. In so far as these traverse little-developed areas any opportunities which they offer for new economic activities should be siezed upon, especially where it is declared government policy to reduce inequalities between the richer and poorer parts of a country. Their main benefits, however, are likely to accrue to the relatively well-developed areas, and especially the towns, which they are linking together. These benefits are even more difficult to assess than those likely to result from new lines penetrating undeveloped areas, but an attempt to investigate these must be made in each case, as must a comparison, in terms of both costs and benefits, with the alternative of improved road connections.

Generally, the assessment of whether the building of new railways represents the best use of scarce resources is no doubt largely the task of the transport economist. But a geographer might be permitted to observe that there is no single answer for the whole of tropical Africa to those who ask whether the era of railway building is now, or should now be, over. The circumstances differ so greatly from place to place that the decision as to whether new rail construction is justified must be made in each case in the light of these.

SELECT BIBLIOGRAPHY

BARBOUR, K. M. (1967). A survey of the Bornu railway extension in Nigeria, *Nigerian Geogr. J.*, **10**, 11–28.

BEST, A. C. G. (1966). *The Swaziland Railway*, Michigan State University Publications, East Lansing.

BILLARD, P. (1966). On construit des chemins de fer au Cameroun, *Revue de Géogr. Alpine*, **54**, 612–20.

FROMM, G. (ed.) (1965). *Transport Investment and Economic Development*, Washington.

HANCE, W. A. (1967). Transport in tropical Africa, in *African Economic Development*, chap. 5, New York and London.

HOYLE, B. S. (1970). Transport and economic growth in developing countries: the case of East Africa, in *Geographical Essays in Honour of K. C. Edwards*, University of Nottingham. See pages 50–62 of this volume.

HUYBRECHTS, A. (1970). *Transports et Structures de Développement au Congo*, Paris.

NICOLAI, H. and JACQUES, J. (1954). 'La transformation des paysages congolais par le chemin de fer: l'exemple du B.C.K.' *Mémoires, Institut Royal Colonial Belge*, pp. 7–208.

NIGERIAN RAILWAY CORPORATION, (1955). *Traffic survey in Zaria, Bauchi and Bornu Provinces*, Lagos.

O'CONNOR, A. M. (1966). *Railways and Development in Uganda*, Nairobi.

O'CONNOR, A. M. (1971). The role of transport, in *The Geography of Tropical African Development*, chap. 7, Oxford.

TAAFFE, E. J. *et al.* (1963). Transport expansion in underdeveloped countries, *Geogr. Rev.*, **53**, 503–29. See pages 32–49 of this volume.

UGANDA PROTECTORATE, (1951). *The Way to the West*, Entebbe.

UGANDA PROTECTORATE, (1956). *Northern Communications*, Entebbe.

Each year, one issue of *Industries et Travaux d'Outre Mer* is devoted to African railways, for example: 'Les grands projets ferroviaires en Afrique', **163**, 500–65, 1967; and 'Les chemins de fer africains et malgaches', **217**, 993–1128, 1971.

9 Container Potential of West African Ports

DAVID HILLING†

MARITIME transport technology has been characterised in recent years by two main trends—the ever-increasing size of bulk carriers and the adoption of improved methods of cargo handling. Improvements in cargo handling have largely revolved around 'unitisation', the process of combining individual packages into large units which can be handled as one. This can make for the quicker and more efficient dispatch of vessels and for a rapid, safe and more economic movement of goods. Unitisation can be achieved by strapping items together (for example, timber, cartons) or placing them on rigid platforms such as pallets or larger 'Lancashire flats' and in each case handling can be by means of fork-lift trucks. It is, however, the container that has attracted most attention and it has become usual to refer to the 'container revolution' that is now in progress. The container, or box, may be of any size, but full advantages of inter-modal transfer only derive from the use of the standard modules which have an 8 ft by 8 ft end section and lengths of 10, 20, 30 or 40 ft. Such containers also have standardised corner members and lifting-points.

Only five years ago the container was an act of faith, but it can now be considered as a fact of life. Few now doubt the ultimate advantages of containerisation and the 'through transport' concept which it makes possible, but it does not follow that implementation can be either universal or everywhere rapid. As the writer has emphasised elsewhere (Hilling, 1967), many of the generalisations on the subject of unitisation, and containerisation in particular, have little universal validity and particular ports and trades have their individual problems. Knowledge and experience of unitised methods of cargo handling are accumulating rapidly, but each new situation requires adaptations and modifications of existing equipment and methods.

The less-developed countries lack adequate capital resources and cannot afford costly experimentation in the sophisticated hardware

† This is a modified and up-dated version of Mr Hilling's article, published in 1969.

of container transport. There is a clear need to narrow the transport gap between the 'haves' and the 'have nots', but there is also a growing amount of evidence which suggests that the unthinking transfer of technology from the socio-economic environment of an advanced economy to the very different environment in a developing country may do little to advance, and indeed could well retard, economic progress. The developing country must avoid the 'two diseases of containeritis . . . a fever of over-enthusiasm and a lassitude of bored or disbelieving reaction' (Crichton, 1969). The progress of container transport in the last five years has been rapid, but it is doubtful if enough is really known of the true economic impact of the new methods for us to join with those who think it 'surprising to see such a marked reluctance on the part of many of these nations to adopt the same enthusiastic approach to the container concept' (Williams, 1969). Why should they? It has yet to be demonstrated satisfactorily that the sophisticated road-rail-sea carrier equipment now being developed for container movement has any immediate relevance to the conditions in many of the developing countries in which the infrastructure is poorly developed, where there is a basic need for roads and railways and where human porterage is in many cases a main form of transport.

Many of the developing countries are characterised by economic and spatial dualism; small islands of modernity are surrounded by a sea of traditional activity, and development economists are by no means unanimous in thinking that this state of affairs is healthy or contributes in the long term to the economic advance of the nation as a whole. The adoption of container transport may well narrow the external transport gap, but it can only serve to exaggerate the internal differences. This article does not attempt to provide an exhaustive feasibility study for container movements in West Africa but rather to make a general appraisal of the prospects. The note sounded is intended to be realistic rather than pessimistic.

WEST AFRICAN TRADE

The area under consideration extends from Cameroon in the east to Mauritania in the west, a total distance of 3,000 km, and embraces sixteen territories and over twenty major ports. Three of the territories are land-locked and dependent on their coastal neighbours for transit traffic (Fig. 9.1). Despite considerable variations, all the

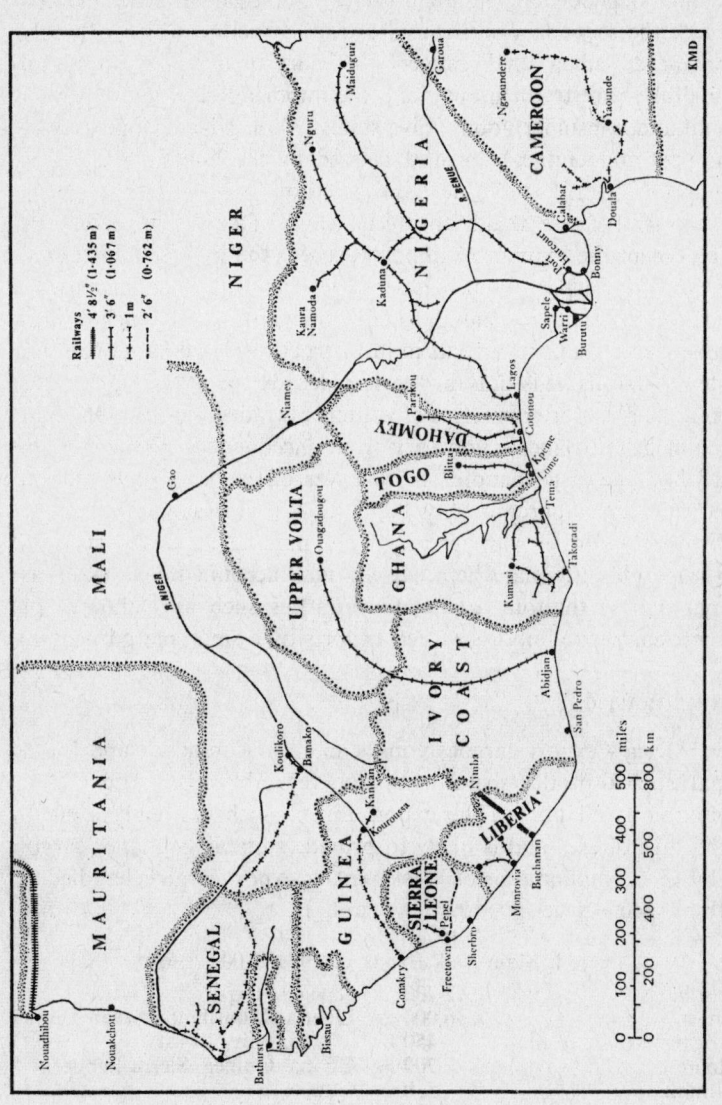

Fig. 9.1 Main ports of West Africa

countries concerned are characterised by low *per capita* incomes, overwhelming dependence on agriculture for employment, little development of the secondary sector and a trade structure which is highly imbalanced with a small range of primary exports deriving from agriculture, forestry or mining and the importation of highly varied capital and consumer goods. Five states (Mali, Niger, Upper Volta, Dahomey and Guinea) are now classed by the United Nations as amongst the world's 'least developed' nations.

It is often the case, that in the initial stages of a country's economic development the tonnage of imports exceeds the tonnage of exports. The growth of export tonnage is dependent on the gradual increase in productive capacity in mining, forestry and agriculture which in turn is dependent on improvements in infrastructure based largely on imported materials. It is only in the last decade that West Africa has become a net exporter in terms of volume of trade (Tresselt, 1967) but the tonnage imbalance is now very pronounced indeed. For the region as a whole the cargo handled in 1969 was of the order of 95 million tons, of which approximately 80 million tons was export cargo. Individual countries vary greatly from mineral exporters such as Liberia or Mauritania where exports may account for as much as 95 per cent of the total traffic, to countries such as Dahomey or Cameroon, where imports exceed exports by a small margin.

EXPORT TRAFFIC

West Africa's export cargoes consist mainly of minerals and traditional tropical products such as timber, vegetable oils, coffee, cocoa, groundnuts and fibres. The export mix is such that only a small percentage of the total is likely to provide suitable container cargo. Well over 73 million tons of West Africa's export cargo is handled in bulk at specially designed berths (Table I).

Table I. Main Bulk Exports—1969 (1,000s tons)

Phosphates	2,700	Senegal, Togo
Iron ore	36,000	Liberia, Mauritania, Sierra Leone
Manganese ore	450	Ghana, Ivory Coast
Bauxite	700	Ghana, Guinea, Sierra Leone
Alumina	530	Guinea
Crude oil and petroleum products	33,000	Nigeria, Ghana

For these cargoes, containerisation offers no advantages from either a technical or an economic point of view.

In the years since World War II there has also been a spectacular increase in timber production and the industry has had a 7·2 per cent per annum growth rate. The bulk of the timber is exported in log form, which precludes the use of containers, while sawn timber, ply and veneers, although increasing greatly in tonnage as industrialisation proceeds, are likely to be unitised in forms other than containers (for example, packaged). Recent suggestions that far more of the timber trade could be handled in full loads on chartered vessels or as bulk loads in liners (UNCTAD, 1970) would further detract from the advantages of containerisation. Log exports of the region now amount to $2\frac{1}{2}$ million tons a year and processed timber amounts to $\frac{1}{2}$ million tons. In the event of conventional rather than cellular vessels being used for containers, some of the ships operating on the West African coast have large hatches and heavy lift-gear for timber handling which would also assist in the handling and stowage of containers either below deck or on the hatch covers.

The third main group of exports comprises agricultural crops. There is still considerable debate as to the extent to which this group of cargoes may be handled in containers. The present transport of groundnuts, palm-kernels, cocoa, coffee, cotton seed, kola, shea and copra is mainly in bags, although on occasions when large shipments are made (for example, groundnuts from Dakar) the produce may be handled in bulk. There is an increasing tendency to do the initial processing of these products before export and in this case the resultant oils and fats are usually shipped in drums or the deep tanks of vessels. Where ships are equipped with deep tanks for such oil products, the desirability of utilising the space may mean that rates will always be more than competitive with charges for container handling (Mittendorf, 1967). The cake by-product of some of these processes may well offer more scope for the use of containers but is easily handled in bag form. Some of the fibres and rubber offer better possibilities, but palletisation may be a more economical proposition in the initial stages.

There have been some trial shipments of cocoa and groundnuts by small container and the view that there would be an unacceptable level of 'sweating' seems not to have been supported. Certainly, it would not be economic to provide specially adapted containers at great cost for one leg of the round voyage, particularly as these products are not of exceptionally high value in relation to weight. Further, these homogeneous cargoes in relatively large consignments are handled fairly efficiently by present methods and containers would not offer any great advantage, especially since the cost of labour at these ports is still relatively low by world standards (Molenaar, 1967). The introduction at Tema of mechanical ship-loaders for handling bagged cocoa and the almost universal use at other ports of bag stackers, mobile conveyors and pre-slinging helps to expedite the handling of bagged cargo. It is interesting to note that an important factor operating against containerisation of these cargoes may well be the usual off-loading at European ports into lighters. At Tilbury 75 per cent of such cargoes move on in lighters, and containers would not be welcome. At a number of the West African ports these agricultural exports make up a high percentage of the northbound traffic (Lagos, 89 per cent; Port Harcourt, 80 per cent; Tema, 70 per cent) and certainly a high percentage of the potential container traffic and may prove to be the critical sector in any container feasibility study. In any year, these exports amount to over 4 million tons.

IMPORT CARGOES

In comparison with the exports, the cargo mix for imports is highly varied and includes a wide range of capital and consumer goods. The import substitution industries on which most developing countries seem to place so much faith are beginning to develop in West Africa but as yet have done little to modify the overall structure of trade except that consumer goods are replaced by intermediate and producer goods. Constructional materials are invariably high on the import list as far as tonnage is concerned and cement frequently takes first place. The import structure of the port of Tema is fairly typical of developing countries (Table II) and reflects the importance of building-up the social and economic infrastructure and the growing need to import food as local agriculture concentrates on export crop production and as urbanisation proceeds.

Table II. Tema—Structure of Imports

	1965	1966	1967	1968
All building materials	57·7	46·8	46·8	37·5
(Cement)	(44·2)	(40·3)	(43·6)	(36·2)
Foodstuffs	15·3	18·9	22·5	24·6
Other	27·0	34·3	30·7	37·9

(percentages of import tonnage)

The quantity of cement may be such that bulk handling becomes feasible, as at Apapa for the cement needed for the construction of the Kainji dam. Cement is often one of the first import substitution industries to be established and in most West African countries is based not on local limestone but on imported clinker. Thus at Tema, Takoradi and Abidjan clinker has largely replaced bagged cement on the import list and is of course handled in bulk. A possible container cargo is thereby eliminated. Vehicles and machinery are often of importance but until there are substantial local assembly industries the demand for container handling will remain slight. For new vehicles there is now a regular roll-on/roll-off service operating between Britain and West Africa.

BALANCE OF TRADE

While exact balance of trade in terms of quantity or quality of cargo is not essential for containerisation, it is clearly desirable to reduce the number of empty containers on return journeys. From what has been said above it is clear that the trade of West Africa, and of most less-developed areas, is seriously imbalanced both in terms of volume and also the nature of the goods. The degree of imbalance varies considerably for different parts of the foreland. To take the case of West Africa's trade with Europe for the year 1965:

Table III. West Africa's Trade with Europe—1965
(1,000s metric tons)

	% of total trade	Exports†	Imports	Northbound surplus
U.K./Ireland	23·5	1,370	988	382
Continent	45·6	2,806	1,756	1,050
French Atlantic	22·6	1,170	1,109	61
Scandinavia	5·9	302	293	9
Baltic	2·4	46	192	−146

† Main agricultural products plus timber.
(from Tresselt.)

These obvious imbalances while not preventing containerisation will certainly detract from the economic advantages to be gained. The balance in the Scandinavian traffic was apparent rather than real, since it resulted from southbound shipment of cement in specialised non-scheduled vessels.

A further interesting feature of the West African trade is its marked seasonal fluctuation. The extent to which fixed growth and harvest times of main agricultural exports result in a cyclic demand for shipping space will depend on the perishability of the product, the urgency of demand and the storage facilities available. It must also be remembered that the internal movement of goods is also determined by climatic conditions, since few West African roads are truly all-season and the advent of the rains is invariably a signal for widespread disruption of the transport system. Figures 9.2 and 9.3 illustrate the degree of variation in traffic from month to month.

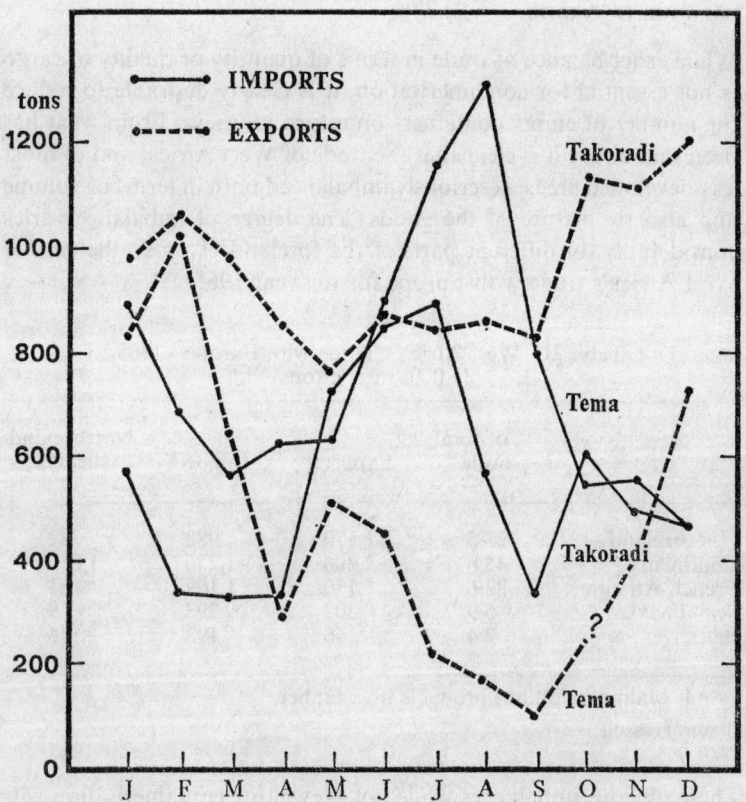

Fig. 9.2   Average tonnage handled per vessel at Ghanaian ports, 1967

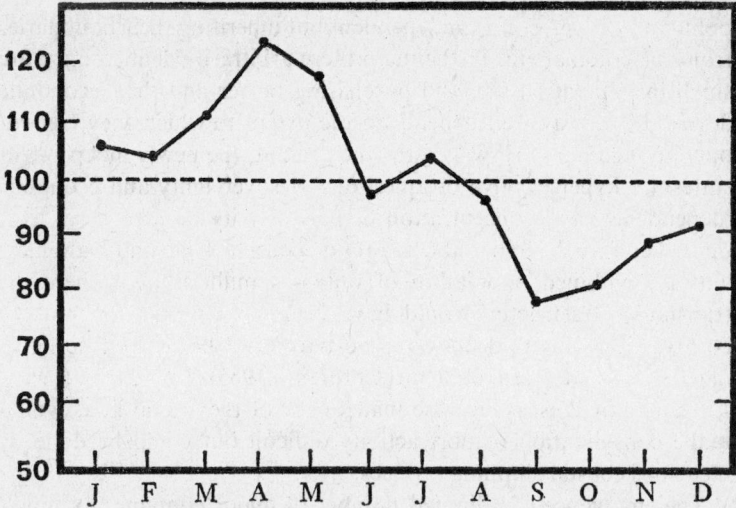

Fig. 9.3   Index of seasonal fluctuations in West African trade (after Tresselt)

The overall potential for containerisation may also be influenced by other characteristics of the trade, namely the tonnages discharged at each port by each vessel and the size of individual consignments. While groupage facilities could take care of the latter the former could provide problems, the full economies of container operation deriving from large throughputs. Figure 9.2 indicates that the average tonnages loaded or unloaded per ship at a port such as Tema are really very small but like all averages conceal considerable variations. Thus in February the average of 681 tons unloaded per ship at Tema represented a range from one ton to 15,650 tons! There has been some consolidation in recent years, but in relation to its volume West Africa's trade is distributed among too many ports; ships tend to call at many of the ports on each voyage since no one port is able to provide anything like a full cargo. This is a severe disadvantage even from the point of view of conventional ship operation but the economics of containerisation demand that ports of call be reduced to a minimum and preferably to one main port at each end of the voyage.

THE POLITICAL FACTOR

The division of trade between so many ports is a result of the political fragmentation of the region. West Africa is divided into many small political units, most now independent but inheriting their boundaries from the colonial era. In the main these arbitrarily defined units are small in population size and purchasing power and their economic development may well depend on the extent to which they can co-operate regionally. However, for the present, the newly independent states are hypersensitive on questions of sovereignty and economic 'dependence' and concentration of port activity on a regional level may, therefore, be impossible to attain. Thus, in Togo and Dahomey, with a combined population of only 4·6 millions and where the transport infrastructure could have been improved to permit the construction of one deep-water port to serve both countries, two new ports have been built, at Cotonou (1965) and Lomé (1968). Elsewhere in West Africa the inadequacy of the lateral links would make consolidation of port activity difficult but could be done by means of coastal shipping services.

The emergence of a limited number of major container terminals possibly feeding more numerous subsidiary ports seems an inevitable trend if container traffic is to develop. On some routes it may be relatively easy to identify pivotal points about which the feeder network could develop, but it is extremely difficult on a linear coast such as West Africa. From what has been said above, it is also doubtful if the states would be prepared to be 'fed' from a major port in some other country and most recent African supranational groupings have been so ephemeral that there is considerable justification for caution in depending on neighbours.

THE TRANSPORT INFRASTRUCTURE

There can be no doubt that the landward links from many of West Africa's ports are quite inadequate even for the satisfactory movement of the present small units of cargo, and that this inadequacy will make the use of all but the smallest containers impossible for some time to come. A recent United Nations study (United Nations Economic Commission for Africa, 1971) concluded that the loading gauges of the region's railways would not preclude the carriage of standard containers but that there would have to be considerable modification of rolling-stock before this could be considered. Additionally, facilities would have to be provided at inland termini for off loading containers and this could only be done at great cost and at a few points. Little of West Africa's road mileage is surfaced (27,000 km out of a total 290,000 km) and many of the roads are open only seasonally. Few of the roads, bridges and culverts could take regular loads in the 20 tons plus range without very considerable up-grading.

Much of West Africa's economic activity is concentrated along a small number of main arteries and development of container transport would emphasise this. However, it is not necessarily in the best interests of economic development that such 'high priority' routes should receive the bulk of investment. The landlocked states of Mali, Niger and Upper Volta should certainly find that their transit traffic through coastal states was improved by use of containers under customs seal. Even so, it would be difficult to justify for three of the world's 'least-developed' and poorest nations any expenditure on container developments at the present time or even, indeed, for a very long time to come.

WEST AFRICAN EXPERIENCE

'The scope for the early development of containerisation in Africa is limited, at any rate in the short term' (Awad, 1967). At the ports of the region there is a clear awareness of the potential benefits of containerisation, given that the conditions permitted implementation. Most of the port authorities accept that the world trend toward containerisation cannot be ignored but there is a general feeling that the situation in West Africa is not yet ready for containers. As import substitution industries develop there may be a time during which consumer goods imports (often suitable for containerisation) decline and capital goods imports (often less suitable) increase. The evidence of advanced countries suggests that the stage of 'high mass consumption' with balanced flows of manufactured goods provides the optimum conditions for container transport. Clearly, this stage of economic growth is still far in the future for West Africa and its attainment could be further delayed if there is large-scale allocation of scarce resources to the implementation of containerisation rather than to productive economic enterprises. This would represent a clear case of what has been called the 'negative' effect of transport provision (Gauthier, 1970).

At the Apapa quay extension at Lagos, one of the berths has been equipped with a 1929 vintage 25-ton crane and has five acres of open back-up area. This could claim to be West Africa's first tailor-made container berth but it has been described as a 'facility lying fallow'. The provision of this berth demonstrated the willingness of the Nigerian Ports Authority to commit itself to containerisation but there is a general feeling of disappointment that the berth, although providing useful space, is not greatly used for containers. African Container Express, a joint enterprise by three shipping companies two British and one Nigerian shipping company, now has over 1,000 10-ton containers in use on the West African coast. Total container movements increased from 800 in 1968 to 2,800 in 1970 and in 1972 should reach 5,000. These containers are carried on conventional ships and for the most part are off-loaded at the normal berths. Nigeria, with a population of 65 millions, may well provide considerable potential for container traffic but few of the small containers now is use move far beyond the immediate port area. Weight limit restrictions would preclude the use of 20-ton containers on most routes and there is presently no suitable handling equipment at interior terminals. Most lorries in use have rigid side walls and could not take containers. There is also the serious disadvantage that the journeys for containers to and from inland locations might have to be measured in weeks or even months. At the present time most of the containers return empty from West Africa, and, while this may be feasible with the present small numbers of containers involved, it could be a serious impediment to the further expansion of operations.

At ports such as Abidjan and Douala some small containers are handled, mainly with southbound consumer-goods cargo; but as in the case of Lagos, they do not penetrate far beyond the port. At Douala studies have been undertaken with a view to the accommodation of larger vessels but the space needed for container operations would probably require the development of Bonaberi, across the river from the main port area. At Douala the port authority have expressed greater interest in palletisation and a long-term preference for LASH (Lighter Aboard Ship) movements. However, while the Niger delta ports of Warri, Burutu and Sapele, with their well-developed river links with the interior, would seem to provide suitable conditions for LASH, as might Abidjan with its lagoons, Douala does not offer the same possibilities.

Cotonou handles part of the traffic of landlocked Niger and such transit movements could clearly benefit from containerisation. However, the transport infrastructure is such that through movement is now impossible. The railway could not now accept even 10 ft ISO containers and the road northwards to Niger from the railhead at Parakou is bituminised to only 3·5 m in width with shoulders that are far from hard, especially in the rainy season. The limited number of small units for the town itself would not seem to justify special facilities. At the recently completed port of Lomé (Hilling, 1969) the establishment of a free zone is designed to attract transit traffic to and from inland neighbours. The total traffic of the port would not justify container facilities at present and the infrastructure, as in the case of Cotonou, would not permit movements inland at the present time.

In Ghana the port authority considered but decided against the provision of container facilities. With axle loads of 14 tons and a loading gauge of 8 ft 4 in., the railways could just handle smaller containers but it was felt that the almost complete lack of suitable return cargo would militate against the use of containers. While Ghana is one of the more industrialised of the West African countries, its exports of manufactured goods are not suitable as return cargo to more advanced economies but may rather be shipped in small amounts to less-developed neighbours. Lake Volta now provides a 400 km artery to the north of the country but the need for transshipment and handling facilities makes the through movement of containers unlikely.

At both Abidjan and Dakar with rail connections of metre gauge to the interior there is transit traffic which would derive benefit from container handling. At both ports there is ample space for development and either could be envisaged as a major container port in the future. As yet, however, movements are largely of non-standard units. At Buchanan, the Liberian port built by the Liberian-American-Swedish Minerals Company (LAMCO) primarily for exporting its iron ore from Mount Nimba, there is a small commercial quay and considerable interest in the use of containers. LAMCO would like to see an increase in container movements of its own general traffic and, with the standard-gauge railway and equipment, containers could easily be handled at Buchanan and also at Yekepa, 265 km up-country. Some small containers have been used for personal effects and for general cargo after groupage at the Farrell Lines pier in New York. At one time, the American West African Conference imposed a heavy-lift surcharge on any unit over two tons. This clearly made nonsense of any attempts to use containers and the surcharge was later withdrawn.

CONCLUSIONS

West Africa is not ripe for full containerisation and is still a long way from partial unitisation, but this is not to argue that these developments will never take place or that moves should not be made in these directions. What is important is that the less-developed countries, faced with a new technology, must weigh very carefully the advantages and disadvantages. Where any new technology is introduced prematurely it may easily lead to disillusionment and even reaction and thereby hinder its eventual acceptance. Since the trade of many of the less-developed countries is either insufficient in volume or unsuitable in kind (although improvements in container equipment could alter this) frequent full container services may be out of the question for a long time to come. Further, the fact that the trade is imbalanced in so many respects (quantity, quality, seasonally) does not favour containerisation which only operates economically with regular, high volume, balanced flows of traffic. Such flows are essential in view of the vast investment necessary for ships, containers and handling equipment. By their very nature, the unit methods of cargo handling in general, and containerisation in particular, are capital- rather than labour-intensive; it can be argued that it is nonsense to substitute scarce capital for labour where there is a large under-utilised labour force and where the cost of labour is low by world standards. Where stevedoring costs are low it is possible that the savings effected by containerisation may be inadequate to justify the capital investment needed (Molenaar, 1967).

While it may be demonstrable that containerisation has considerable advantages for the ship-owner the economic advantages for the less-developed country are certainly not clear. Such countries should therefore heed Nagorski who has written that 'over-ambitious projects should be avoided and building new port facilities should be arranged cautiously, on the basis of sober traffic estimates rather than loose hopes' (Nagorski, 1968). These words of caution could perhaps be adopted in other areas of transport provision. Yet, undue caution could lead to an ever-widening transport 'gap' between the advanced and less-advanced economies. Programmes have therefore been suggested (Williams, 1969) for the gradual implementation of container methods, and, while these may provide useful guidelines, the less-developed country would do well to explore thoroughly all aspects of what might be termed 'intermediate technology' (pallets, cargo cages, small collapsible containers, etc). Pallets in particular could greatly expedite cargo handling without great capital investment. Since none of these methods of handling ultimately exclude containerisation, nothing would be lost and much may be gained in the meantime. Perhaps of greater importance, with LASH operations now providing regular services between American Gulf ports and Europe and more routes about to be established, the less-developed country leaves open a valuable option.

In conditions such as exist in many less-developed countries, it may well be that the inter-modal transportation system represented by LASH will have more to recommend it than containerisation. Whereas containerisation is bound to involve costly modifications or additions to port facilities the LASH ship merely requires a large mooring buoy in sheltered water. Having off-loaded its barges, the ship can proceed on its way without costly delays in port and the barges can be unloaded at existing berths, even in shallow water, using the most suitable methods. The barges may in some situations, and ideally, be moved on by water—this might be possible on the Gambia and Niger/Benue rivers or on the lagoons of southern Ivory Coast, Dahomey and Nigeria. LASH ships would seem not to be at any great disadvantage when having to call at a number of ports but this could well eliminate containerisation except by way of a pivotal port and feeder services with the political problems that this would create. LASH barges could possibly be used for the coastwise consolidation of cargo which inadequate land transportation now often precludes and so could reduce the number of port calls by main-line vessels, a major factor in economic ship operation.

Some of the main shipping companies operating to West Africa are known to be considering containerisation but it is thought that a full container service with cellular vessels would cost about £30 million. Given the nature of the West African trade, it seems most unlikely that a shipping company could expect any substantial returns on investment of this order, and as a consequence the chances of a less-developed country benefiting from reduced freight rates seem remote; and a less-developed country would in any case have to balance any saving from reduced freight rates against the cost of providing handling facilities. Until much more is known of the cost benefits of containerisation it would be unwise for the countries of West Africa to invest large sums of money in this new technology.

REFERENCES

AWAD, S. (1967). African experience, *Proceedings of United Nations Interregional Seminar on Containerisation* (May), London.

CRICHTON, A. (1969). The container revolution arrives, *The Journal of Commerce Annual Review*, pp. 33–7.

GAUTHIER, H. L. (1970). Geography, transport and regional development, *Economic Geography*, **4**, 612–19. See pages 19–31 of this volume.

HILLING, D. (1967). Report on United Nations interregional seminar on containerisation, *The Dock and Harbour Authority* (August), pp. 121–4.

HILLING, D. (1969). Togoport—new port of Lomé, *The Dock and Harbour Authority* (March), pp. 423–4.

MITTENDORF, H. J. (1967). Some considerations on containerisation in marketing agricultural products, *Proceedings of United Nations Interregional Seminar on Containerisation* (May), London.

MOLENAAR, H. J. (1967). Basic economics of containerisation and unitisation in ocean shipping, *Proceedings of United Nations Interregional Seminar on Containerisation* (May), London.

NAGORSKI, B. (1968). Port problems in developing countries, *The Dock and Harbour Authority* (June), pp. 36–43.

TRESSELT, D. (1967). *The West African Shipping Range*, United Nations, New York.

UNCTAD (1970). *The Maritime Transportation of Tropical Timber*, TD/B/C.4/59.

UNITED NATIONS ECONOMIC COMMISSION FOR AFRICA (1971). *Shipping and Ports—Suggested African positions for Third Session of UNCTAD*, Geneva.

WILLIAMS, G. R. (1969). Containerisation in developing countries, *The Dock and Harbour Authority* (January), pp. 348–52.

10 Transportation and the Growth of the São Paulo Economy

HOWARD L. GAUTHIER

INTRODUCTION

THIS study is concerned with the general problem of transportation development and economic change. Specifically, it examines the development of highway transportation and urban growth in the region of São Paulo, Brazil, an area that is undergoing a rapid economic transformation. The objective of the analysis is to investigate, for the period 1940 to 1960, the interrelationships between changes in accessibility to the highway network and the growth of urban centres.

Basic to this study is the notion that there exists a high degree of interdependence between the development of a transportation system and the geographic pattern of urban economic growth. One may consider capital investments that lead to additions and changes in a transportation network as shocks that are felt throughout the entire transport system. One possible consequence of these shocks may be an alteration in the spatial structure of the network. The change in network structure may have an effect on economic development. It can produce serious changes in the pattern of internal accessibility for many of the urban centres of the network. In turn, the change in accessibility may disrupt existing patterns of spatial competition in the region and have an effect on relative rates of urban growth.

NETWORK STRUCTURE AND ACCESSIBILITY

A consideration of the interrelationships between changes in accessibility to a transportation network and the growth of urban centres requires that the properties of network structure be handled in some precise manner. The term structure denotes the spatial configuration or pattern of a network and includes properties of the component elements of the network that are related to this configuration. To obtain this precision, the São Paulo highway network is treated as a representation of a finite graph. Viewed from the perspective of graph theory, a highway network is a set of vertices (nodes) which are related by an incidence function according to a configuration of arcs (linkages). As a representation of a finite graph the São Paulo highway network is a planar, symmetrical, and connected graph. (See Figs. 10.1, 10.2 and 10.3.)

To reduce the loss of empirical information that is inevitable when a network is generalised as an abstract graph, a highway network can be considered in terms of a valued graph. By assigning values to the arcs, it is possible to distinguish node-linkage associations according to relations of varying strength and intensity. In this study, the arcs are valued by transfer costs in relation to length of haul. The purpose of the valuation is to differentiate node-linkage associations according to : (1) the distance between the urban nodes, (2) the type of connecting linkage, for example, paved, graded, or laterite, and (3) changes in transfer cost gradients resulting from improvements in the mode of transfer.

The analysis of transfer costs as a function of distance has received little attention in studies of transportation in Brazil. Fortunately, the few studies that are available do permit us to make estimates for the São Paulo network. The estimating procedure commonly employed by Brazilian transportation economists and traffic engineers is the one suggested by Pires Ferreira (1940). In his analysis of transfer costs in relation to length of haul, Sr. Pires sets forth the formulation,

$$y^2 - D_1^2 x^2 - 2D_1 D_2 x - 2k = 0 \qquad (1)$$

where

$y =$ the total costs
$D_1 =$ the variable costs incurred by a vehicle travelling a distance x
$D_2 =$ the fixed overhead charges
$2k =$ initial outlays for equipment and administrative expenses; considered equal to the constant of integration multiplied by 2.

Allowing $C = -D_1^2$, $E = -(D_1 D_2)$, and $F = -2k$, equation (1) can be rewritten as,

$$y^2 + Cx^2 + 2Ex + F = 0 \qquad (2)$$

Thus, we have a special case of the general second-degree equation,

$$Ay^2 + 2Bxy + Cx^2 + 2Dy + 2Ex + F = 0 \qquad (3)$$

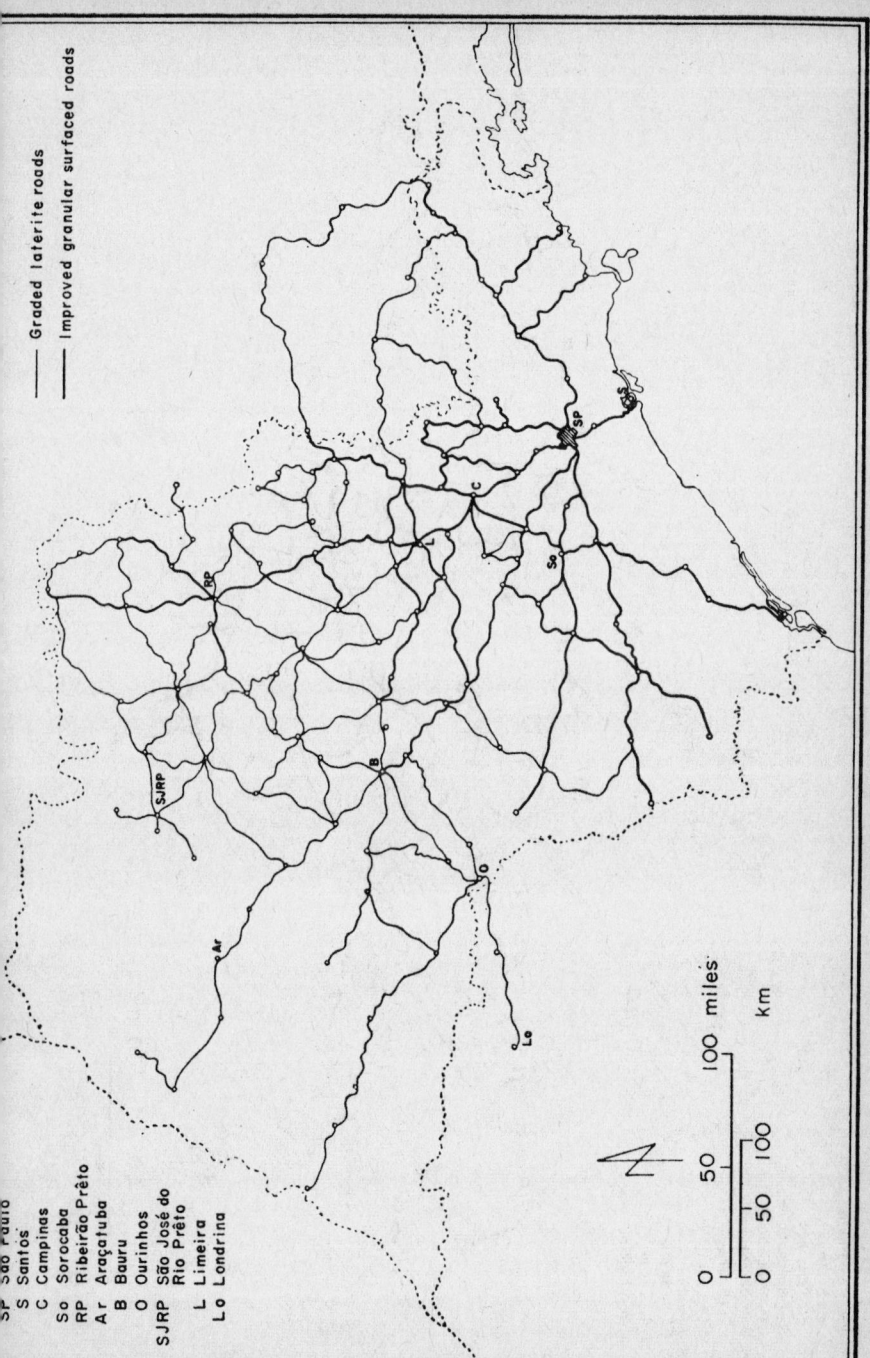

Graded laterite roads

Improved granular surfaced roads

SP São Paulo
S   Santos
C   Campinas
So Sorocaba
RP Ribeirão Prêto
Ar Araçatuba
B   Bauru
O   Ourinhos
SJRP São José do Rio Prêto
L   Limeira
Lo Londrina

100 miles

km

0   50   100

0   50   100

Fig. 10.1   São Paulo highway network, 1940

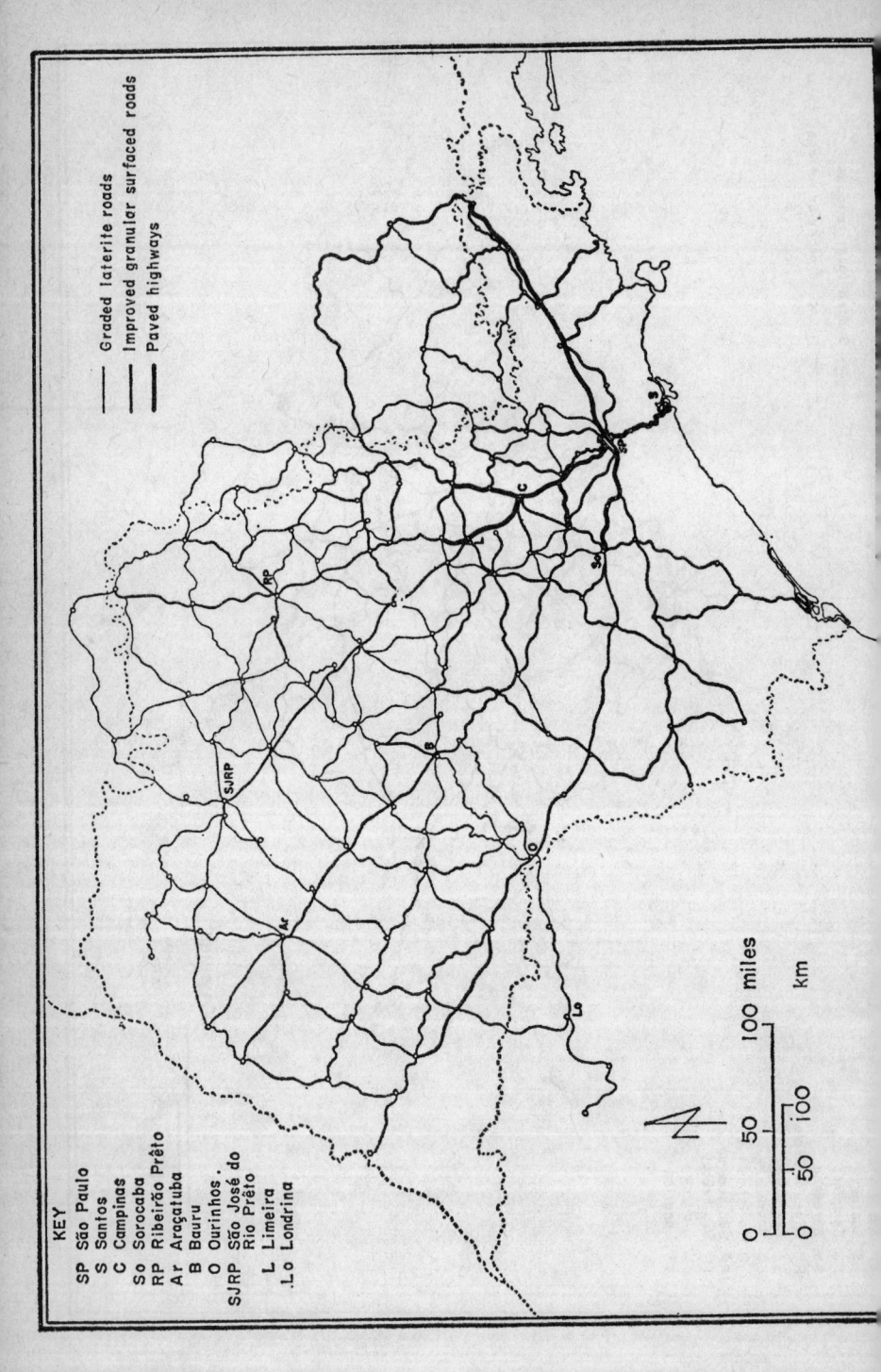

KEY
SP São Paulo
S   Santos
C   Campinas
So Sorocaba
RP Ribeirão Prêto
Ar Araçatuba
B   Bauru
O   Ourinhos .
SJRP São José do
      Rio Prêto
L   Limeira
Lo Londrina

——— Graded laterite roads
——— Improved granular surfaced roads
——— Paved highways

0   50   100 miles
0   50   100
       km

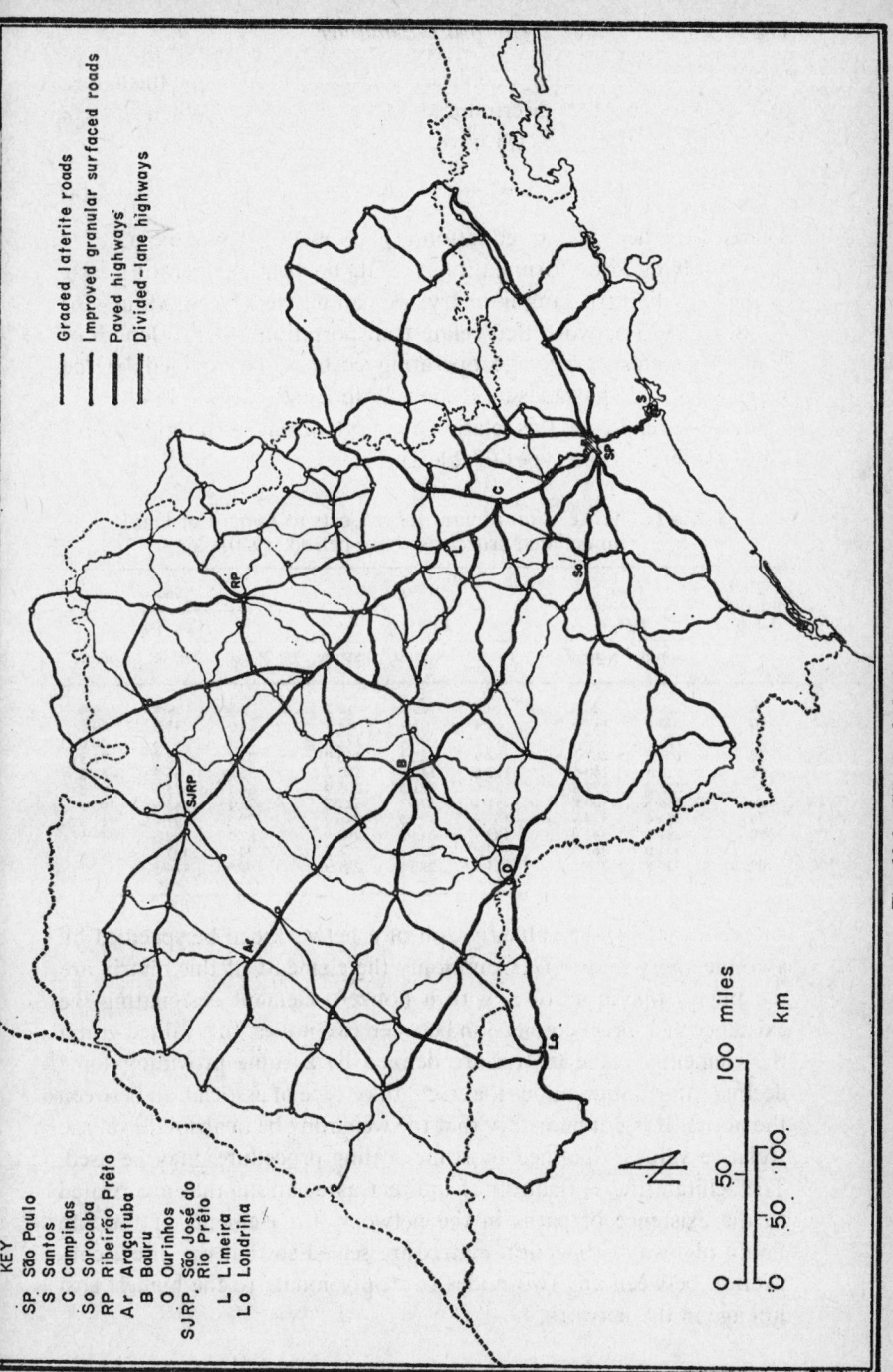

KEY

SP São Paulo
S   Santos
C   Campinas
So Sorocaba
RP Ribeirão Prêto
Ar Araçatuba
B   Bauru
O   Ourinhos
SJRP São José do
     Rio Prêto
L   Limeira
Lo Londrina

—— Graded laterite roads
—— Improved granular surfaced roads
—— Paved highways
—— Divided-lane highways

0   50   100 miles
0   50   100
          km

Fig. 10.3 São Paulo highway network, 1960

In the manner of treating a conic section, we can determine the nature of the curve by the discriminant, $\Delta = B^2 - 4AC$. When $B = 0$, $A = 1$, and $C = -D_1^2$, we have,

$$\Delta = -C = D_1^2 > 0 \qquad (4)$$

Therefore, when $\Delta > 0$, equation (1) represents a hyperbola.

By applying Pires' formulations to data on average operating costs in the Brazilian trucking industry, we can derive a set of values for the São Paulo network that relate transportation costs to length of haul. Estimates of average operating costs were provided by the Departmento de Estradas de Rodagem do Estado de São Paulo. The set of values expresses this relationship both through time and according to different road types (Table I).

Table I. The Relation of Transport Costs to Length of Haul
(in deflated cruzeiros; base period, 1940)

Distance (klms)	1940 Cr$/Ton		1950 Cr$/Ton			1960 Cr$/Ton		
	Gravel	Laterite	Paved	Gravel	Laterite	Paved	Gravel	Laterite
0	26·0	26·0	19·7	19·7	19·7	9·4	9·4	9·4
10	90·3	117·6	54·4	68·4	89·1	25·9	31·8	40·5
20	128·6	168·4	79·2	97·4	127·6	37·7	45·3	58·0
30	161·2	207·1	98·9	122·1	156·9	47·1	56·8	71·3
40	187·8	240·5	117·6	142·3	182·2	56·0	66·2	82·8
50	213·2	271·8	134·6	161·5	205·9	64·1	75·1	93·6
60	236·7	300·6	151·4	179·3	227·7	72·1	83·4	103·5
70	259·4	324·9	167·6	196·5	246·2	79·8	91·4	111·9
80	281·8	347·0	182·9	213·5	262·9	87·1	99·3	119·5
90	304·0	367·6	197·8	230·3	278·5	94·2	107·1	126·6
100	325·8	385·3	211·9	246·8	291·9	100·9	114·8	132·7

Operationally, the valued graph of a network can be specified by a connectivity matrix G. Commonly the elements of this matrix are the binary operators 0, 1, with a nonzero element designating the existence of a direct connection between two nodes. In a valued graph the elements of the matrix are defined by a value providing some desired information about the strength or type of association between the nodes. It is not necessary that the weighting be in absolute values. Relative values, obtained by some scaling procedure, may be used. To facilitate the evaluation of indirect associations that are related to the existence of paths in the network, the elements of the São Paulo highway connection matrix are scaled so that the direct connection between any two nodes is proportionate to the highest cost linkage in the network.

Given n geographical locations x_i, for every pair $(x_i x_j)$ a connectivity index x_{ij} is defined. If $(x_i x_j)$ are not connected, the value of the index is zero. If $(x_i x_j)$ are connected, the index value is greater than zero and less than unity, depending on the length of haul between two centres and the type of road over which a vehicle must travel. The $n \times n$ indices define the connectivity matrix which has in the ith row and jth column the index value y_{ij}, that is, the measure of connectivity between x_i and x_j. The graph corresponding to the matrix has appropriate, positive, decimal loadings for each of its arcs.

Because the order of the connectivity matrix for each time period is in excess of 120×120 it is not feasible to present the data matrices of the entire São Paulo highway network. By way of illustration, Tables II, III and IV show the weighted connectivity matrices for a 15-node subgraph of the highway network in the southeastern section of the study area.

The advantage of considering the São Paulo highway network as a representation of a valued graph is apparent in an appraisal of connectivity at different time intervals. Based on the weighting system employed, changes in the values identifying node-linkage associations result from the impact of one or more of the following factors: (1) a decrease in actual road distances (a not uncommon occurrence when laterite roads are replaced with improved granular-surfaced ones), (2) an upgrading of the structural quality of the linkage, or (3) the introduction of cost-saving innovations, for example, the introduction of the diesel-powered truck. Obviously, the more factors affecting a nodal linkage through time the greater is the change in connectivity.

By a power expansion of the connectivity matrix, it is possible to determine the indirect connectivity between the nodes in the network, that is, a path between x_i and x_j by an intermediary node x_k. The power expansion is terminated when all paths of sequence length n between x_i and x_j have been determined. An element y_{ij}^n of the connection matrix G^n lies within the following bounds $0 \leq y_{ij}^n \leq y_{ij}^{n-1} \leq y_{ij}^{n-(n-1)}$. As is evident from the inequality, the index of connectivity assigns a greater value to a direct connection than to an indirect one. The incremental change in the value of indirect connectivity decreases with an increase in the length of the path sequence. In this manner, it is possible to incorporate into the analysis a distance decay effect on nodal associations.

Table II. Connection Matrix for Subgraph of Highway Network, 1940

Centre	Direct connections														
	1	2	3	4	5	6	7	8	9	10	11	12	13	14	15
1. São Paulo	×	0·66		0·55			0·78								
2. Mogi das Cruzes	0·66	×	0·57												
3. São José dos Campos		0·57	×												
4. Atibaia	0·55			×	0·83	0·85									
5. Piracaia				0·83	×										
6. Bragança Paulista				0·85		×									
7. Franco da Roacha	0·78						×	0·78							
8. Jundiaí							0·78	×	0·70						
9. Campinas								0·70	×	0·85	0·75				
10. Jaguariúna									0·85	×					
11. Americana									0·75		×	0·74	0·86		
12. Piracicaba											0·74	×	0·77		
13. Limeira											0·86	0·77	×	0·83	0·86
14. Araras													0·83	×	0·84
15. Rio Claro													0·86	0·84	×

Table III. Connection Matrix for Subgraph of Highway Network, 1950

Centre	Direct connections														
	1	2	3	4	5	6	7	8	9	10	11	12	13	14	15
1. São Paulo	×	0·71	0·57	0·61	0·51		0·78	0·74							
2. Mogi das Cruzes	0·71	×	0·63												
3. São José dos Campos	0·57	0·63	×												
4. Atibaia	0·61			×	0·85	0·87		0·72							
5. Piracaia	0·51			0·85	×	0·71									
6. Bragança Paulista				0·87	0·71	×									
7. Franco da Rocha	0·78						×	0·80							
8. Jundiaí	0·74			0·72			0·80	×	0·82						
9. Campinas								0·82	×	0·90	0·83				
10. Jaguariúna									0·90	×					
11. Americana									0·83		×	0·78	0·91		
12. Piracicaba											0·78	×	0·80		0·81
13. Limeira											0·91	0·80	×	0·89	0·88
14. Araras													0·89	×	0·86
15. Rio Claro												0·81	0·88	0·86	×

Table IV. Connection Matrix for Subgraph of Highway Network, 1960

Centre	Direct connections														
	1	2	3	4	5	6	7	8	9	10	11	12	13	14	15
1. São Paulo	X	0·81	0·63	0·75	0·57		0·82	0·78							
2. Mogi das Cruzes	0·81	X	0·69												
3. São José dos Campos	0·63	0·69	X												
4. Atibaia	0·75			X	0·88	0·92		0·76							
5. Piracaia	0·57			0·88	X	0·78									
6. Bragança Paulista				0·92	0·78	X									
7. Franco da Rocha	0·82						X	0·84							
8. Jundiaí	0·78			0·76			0·84	X	0·85						
9. Campinas								0·85	X	0·95	0·86				
10. Jaguariúna									0·95	X					
11. Americana									0·86		X	0·89	0·92		
12. Piracicaba											0·89	X	0·83		0·84
13. Limeira											0·92	0·83	X	0·91	0·92
14. Araras													0·91	X	0·88
15. Rio Claro												0·84	0·92	0·88	X

The summation of G, G^2, ..., G^n is a matrix T which can be interpreted as a numerical representation of the accessibility surface of the network. Each element t_{ij} is a measure of direct and indirect connectivity between x_i and x_j. The attenuated relations of indirect connectivity include both elementary and redundant paths. There is a directly proportional relationship between the value of t_{ij} and the accessibility between nodes of the network, that is, the higher the value of t_{ij} the greater the accessibility between x_i and x_j. Variations in the values of the elements of the accessibility surface are the result of: (1) the number of direct linkages incident to a node, (2) the number of open and closed path sequences, and (3) the length and constructional quality of both direct and indirect node-linkage association. The accessibility surfaces of the 15-node sub-graph of the São Paulo highway network are presented in Tables V, VI and VII.

STRUCTURAL DIMENSIONS OF THE ACCESSIBILITY SURFACES

To determine the structural dimensions of the accessibility surfaces of the São Paulo highway network, the T matrices are subjected to a principal axes factor solution, with communality estimates of unity. This multivariate procedure, combined with a Varimax rotation to simple structure, generates a factor matrix in which each factor is described in terms of those variables with which it is most highly correlated. The result is an economy of expression in terms of the number of relevant factors. For the years 1940, 1950 and 1960, the principal axes factor solution, with Varimax rotation, extracted five factors that were relevant. In each successive time period, they accounted for 61·4, 68·0, and 76·7 per cent of the total variance. Because of the extensive amount of data generated by a principal axes factor solution for three time periods, it is not feasible to present here the factor matrices. Copies of these matrices are available from the author.

The five rotated factors result from the collapse of highly correlated connections between nodes of the network into distinctive clusters of nodes which display similar patterns or profiles of connectivity, both direct and attenuated. The analysis, then, 'classifies' nodes on the basis of similarity in their accessibility to the highway network. Each factor represents a linear combination of a group of nodes which forms a basic structural dimension of the regional network.

Table V. Accessibility Surface of Subgraph of Highway Network, 1940

Centre		Accessibility values													
	1	2	3	4	5	6	7	8	9	10	11	12	13	14	15
1. São Paulo	X	26·3	17·0	36·7	29·1	28·2	12·6	40·0	21·7	16·4	10·1	3·4	5·6	3·8	2·4
2. Mogi das Cruzes		X	12·9	16·6	12·2	10·0	6·1	12·3	9·6	4·2	2·4	1·1	1·6	1·7	0·5
3. São José dos Campos			X	3·5	2·8	2·5	1·2	3·3	1·5	1·0	0·6	0·2	0·2	0·1	0·1
4. Atibaia				X	20·7	19·3	11·6	22·5	23·0	10·9	7·2	4·5	7·2	4·4	3·5
5. Piracaia					X	15·0	7·5	18·1	13·5	8·4	5·0	2·4	4·0	2·6	1·8
6. Bragança Paulista						X	6·6	20·9	13·2	11·4	7·3	3·5	6·4	4·8	3·2
7. Franco da Rocha							X	8·4	11·5	4·6	3·7	2·8	4·3	2·6	2·3
8. Jundiaí								X	22·7	20·8	16·1	8·2	14·7	11·3	7·8
9. Campinas									X	20·1	21·8	18·0	30·3	21·9	18·2
10. Jaguariúna										X	20·8	15·3	29·3	24·7	17·4
11. Americana											X	22·7	40·4	32·8	26·1
12. Piracicaba												X	37·6	30·3	25·2
13. Limeira													X	57·2	44·8
14. Araras														X	36·9
15. Rio Claro															X

Table VI. Accessibility Surface of Subgraph of Highway Network, 1950

Centre	Accessibility values														
	1	2	3	4	5	6	7	8	9	10	11	12	13	14	15
1. São Paulo	×	219·7	162·8	339·7	220·7	197·6	281·9	590·6	340·4	148·7	129·1	152·6	758·4	50·9	68·5
2. Mogi das Cruzes		×	59·1	115·8	76·3	64·9	96·7	192·6	103·9	44·1	35·0	37·9	23·4	11·2	15·2
3. São José dos Campos			×	86·5	57·1	48·4	72·3	143·1	76·9	32·6	25·9	27·9	17·3	8·3	11·3
4. Atibaia				×	122·2	110·3	154·3	318·2	193·0	84·4	73·7	89·2	61·0	33·1	43·2
5. Piracaia					×	73·6	98·8	203·4	114·9	53·5	42·3	46·4	32·2	17·5	21·6
6. Bragança Paulista						×	86·3	188·1	107·9	58·9	46·1	50·9	40·9	25·9	28·1
7. Franco da Rocha							×	267·1	164·5	69·3	64·5	80·3	53·4	28·2	38·7
8. Jundiaí								×	304·3	180·6	182·5	233·7	167·6	96·9	127·8
9. Campinas									×	137·4	178·4	262·4	203·7	126·2	167·9
10. Jaguariúna										×	98·9	134·6	123·8	88·0	100·0
11. Americana											×	234·8	210·1	144·3	184·9
12. Piracicaba												×	373·9	327·6	291·9
13. Limeira													×	225·5	276·9
14. Araras														×	199·6
15. Rio Claro															×

Table VII. Accessibility Surface of Subgraph of Highway Network, 1960

Centre	Accessibility values														
	1	2	3	4	5	6	7	8	9	10	11	12	13	14	15
1. São Paulo	X	480·4	363·2	753·9	486·9	473·3	572·3	934·9	697·6	314·5	256·4	312·0	188·5	97·3	133·5
2. Mogi das Cruzes		X	135·9	266·4	173·6	158·9	203·4	415·7	220·9	96·8	72·1	80·9	47·1	22·5	31·0
3. São José dos Campos			X	203·2	132·8	121·1	155·2	314·7	166·1	72·9	53·9	60·1	35·2	16·9	23·1
4. Atibaia				X	282·7	273·9	328·6	690·9	409·0	183·5	149·1	185·1	119·2	64·2	85·0
5. Piracaia					X	182·2	209·1	440·6	242·8	117·1	85·9	97·5	63·9	34·8	43·0
6. Bragança Paulista						X	195·9	442·2	244·2	139·9	100·4	112·4	87·1	55·6	59·4
7. Franco da Rocha							X	530·4	320·8	137·7	120·8	154·9	96·8	50·4	70·9
8. Jundiaí								X	762·6	366·1	350·2	457·3	309·7	175·6	238·9
9. Campinas									X	264·6	337·3	510·2	369·2	223·8	311·4
10. Jaguariúna										X	180·4	247·6	215·9	151·8	175·6
11. Americana											X	438·3	368·6	248·1	333·8
12. Piracicaba												X	586·7	394·2	538·2
13. Limeira													X	377·9	482·1
14. Araras														X	341·7
15. Rio Claro															X

Dimension One

In 1940, the first dimension of the accessibility surface is characterised by high positive loadings on 40 of the region's 123 nodes. As a dimension of the accessibility surface, this factor constitutes the major structural component of the network, accounting for 16·5 per cent of the total variance. Empirically, the dimension may be labelled the São Paulo Maior Region, a term commonly used to identify the São Paulo–Campinas–Sorocaba triangulation and an axial extension along the Upper Paraíba Valley.

The loadings on factor one for 1950 and 1960 reveal a profile of accessibility identical to that of 1940. Among the nodes characterised by significant loadings in the earlier period, it is clearly evident that the degree of identification with the São Paulo Maior Region tends to be strengthened in successive time periods. In addition, there is every indication of an increase in the number of nodes that are identifiable with this dimension of the network. As a result of the increases in both magnitude and number of the loadings of the nodes on factor one, this dimension of the accessibility surface has maintained its primacy through time. Spatially, the expansion of this dimension through time has occurred along the major highway arteries that converge on the metropolitan area of São Paulo.

Dimension Two

Since 1940, there has been an increase in the relative importance of the second dimension of the network, as evidenced by eigenvalues that explain 13 per cent, 14 per cent, and 17 per cent of the original variance in the years 1940, 1950, and 1960. Empirically in all three time periods, the second dimension is distinguished by a cluster of centres that have transportation linkages which have been associated historically with the development of coffee as a principal cash crop in the regional economy. Significant loadings are characteristic of the centre of Riberão Prêto and 35 other centres geographically clustered in the northern part of the State of São Paulo.

Dimension Three

Whereas the first and second dimensions of the structure of the highway network have been identifiable in all time periods with the first and second factors of the rotated factor matrices, the three remaining dimensions are characterised by a change through time in the rank-order of their factorial vectors. The change in rank-order is the result of the increasing importance of one of the major dimensions of the network structure. In 1940 this dimension ranks fifth in the order of the factorial vectors, accounting for only 8·8 per cent of the original variance. In 1950, it ranks fourth and by 1960 has advanced to third position, accounting for 12·5 and 15·5 per cent of the variance.

High positive loadings for the centres of São José do Rio Prêto, Araçatuba, Bauru, and Ourinhos empirically identify the dimension as the Western or Pioneer Region. All of the centres aligned on this dimension have been the recipients of improved connectivity and accessibility to the highway network resulting from the postwar expansion of highway construction. More than 5,000 km have been built in the region since 1940. The construction and upgrading of highway facilities in the Pioneer Region is reflected in the relative increases in explained variability associated with this dimension. In each time period more nodes have been added to this dimension than to any other, including the São Paulo Maior Region.

Dimensions Four and Five

Unlike the first three dimensions, these two have failed to either maintain or improve their relative positions in the rank-order of the factorial vectors. Both dimensions are regional groupings of centres lying between the expanding areas of the São Paulo Maior Region and either the Riberão Prêto Region or the Pioneer Region. Empirically, they may be identified as the Sul de Minas Region and the Southern Region.

According to the principal axes factor solution of network structure in 1960, there are fewer centres aligned on the fourth and fifth dimensions than in either 1940 or 1950. If this trend continues, one might expect a realignment of the nodes comprising these dimensions on the first three dimensions of the network.

NETWORK ACCESSIBILITY AND URBAN GROWTH

The determination of the accessibility surfaces of the São Paulo highway network and the extraction of the structural dimensions of those surfaces provide a basis for answering questions as to the extent changes in network accessibility are related to the economic growth of urban centres. Operationally, the structural dimensions of the accessibility surfaces provide measures of network accessibility at a regional scale. Measures of urban growth are provided by the surrogates of urban population, manufacturing, and retail trade activity. By using the multivariate procedure of canonical correlation, the relationships between the two sets of measures can be determined. The objective is to maximise the covariance between two linear combinations of the two sets of variates.

Following Anderson (1958), let us assume we have p variables, X_1, X_2, X_3, ..., X_p and N observations on each variable. The variables are divided into two groups

$$\left\{ \begin{array}{c} X_1 \\ X_2 \\ \cdot \\ \cdot \\ \cdot \\ X_{p1} \\ \hline X_{p1+1} \\ X_{p1+2} \\ \cdot \\ \cdot \\ \cdot \\ X_{p1+p2} \end{array} \right\} = \left[\begin{array}{c} X^{(1)} \\ \hline X^{(2)} \end{array} \right].$$

As X is partitioned into two subvectors of p_1 and p_2 components, the variance–covariance matrix is partitioned into p_1 and p_2 rows and columns,

$$XX^T = \begin{bmatrix} X^{(1)}X^{(1)} & X^{(1)}X^{(2)} \\ X^{(2)}X^{(1)} & X^{(2)}X^{(2)} \end{bmatrix}$$

with

$$EXX^T = \begin{bmatrix} \Sigma11 & \Sigma12 \\ \hline \Sigma21 & \Sigma22 \end{bmatrix}.$$

A canonical analysis requires linear combinations of the components $X^{(1)}$ and $X^{(2)}$ such that

$U = \alpha^T X^{(1)}$ and $V = \gamma^T X^{(2)}$, with corresponding correlations, $\sqrt{\lambda_1} \geq \sqrt{\lambda_2} \geq \ldots \geq \sqrt{\lambda_p}$. To obtain α and γ,

$$\begin{pmatrix} -\lambda\Sigma11 & \Sigma12 \\ \Sigma21 & -\lambda\Sigma22 \end{pmatrix} \begin{pmatrix} \alpha \\ \gamma \end{pmatrix} = 0.$$

To determine the relationships between network accessibility and urban growth in the São Paulo Region, we have eight variables which can be divided into two component groups, $X^{(1)}$ and $X^{(2)}$. The elements of the first set of variates $X^{(1)}$, are the surrogate measures of economic development

X_1 = Value of industrial production
X_2 = Urban population
X_3 = Value of retail sales.

The elements of the second set of variates, $X^{(2)}$, are the structural dimensions of the highway accessibility surfaces

X_4 = São Paulo Maior Region
X_5 = Riberão Prêto Region
X_6 = Pioneer Region
X_7 = Sul de Minas Region
X_8 = Southern Region.

The results of the canonical analysis for the period 1940 to 1960 are summarised in Table VIII. To interpret the canonical variates, the coefficients of the component variable are considered similar to the loadings in factor analysis. They are evaluated in terms of the magnitude of their values, in combination with the direction of their signs.

Table VIII. Canonical Analysis, 1940–60

Root	$X^{(1)}$			$X^{(2)}$					$\sqrt{\lambda}$	χ^2	D.F.
	X_1	X_2	X_3	X_4	X_5	X_6	X_7	X_8			
1	5·20	2·18	1·54	9·23	4·01	−3·18	1·48	0·72	0·42	58·2‡	15
2	2·42	6·78	1·06	−0·42	−0·32	1·94	0·23	−0·19	0·31	13·8†	8
3	1·19	1·11	−0·95	−0·14	−0·94	−0·02	0·47	0·08	0·08	1·3	3

† Significant at 0·95 confidence level.
‡ Significant at 0·99 confidence level.

Any attempt to interpret the results of the canonical analysis must proceed with a certain amount of caution. Only a limited number of roots have been extracted. This is the result of the consideration of only three variables to represent urban economic development. Certainly other relationships may exist between network accessibility and urban growth in addition to those revealed in this study.

A critical examination of the coefficients of the canonical variates in Table VIII suggests that they have empirical significance. The first pair of variates reveals that the largest weights of the U_1 and V_1 eigenvectors have to be given to increases in the value of manufacturing and accessibility to the São Paulo Maior Region. With respect to the interrelationships between accessibility and urban growth, we may conclude that the strongest association exists between the construction of highway facilities in the São Paulo Maior Region and the development of manufacturing activities in urban centres which have been the beneficiaries of the resulting improvements in network accessibility.

It will be recalled that the São Paulo Maior Region is the major structural components of the network accessibility surfaces in all time periods. Spatially, the limits of the region have increased through time, with additional centres having significant loadings on this dimension. These centres are all aligned along the major arterial routes that converge on the metropolis of São Paulo. Certainly, it is reasonable for the development of manufacturing production to occur in those urban centres having the highest degree of accessibility to the highway network. Presumably, improved accessibility favours the growth of industry and industrial development creates a demand for an increase in transport accessibility.

After the primary relationship between accessibility and urban growth has been extracted, there is a second way in which the sets of variates $X^{(1)}$ and $X^{(2)}$ can be arranged in a significant manner. The coefficients of the canonical variates U_2 and V_2 indicate a relationship between increases in accessibility to the Pioneer Region and a growth in urban population.

It will be recalled that the Pioneer Region has undergone a significant improvement through time in its rank order among the structural dimensions of the highway network. This increase is greater than that of any other dimension extracted from the rotated factor matrices. It is the result of the spatial expansion of the dimension. In each time period, more nodes were added to this dimension than to any other. Concomitant with the increase in network accessibility, the period from 1940 to 1960 was one of: (1) an increasing concentration of population in established centres, and (2) the creation of new centres as services centres for developing agricultural areas.

As it is possible to provide empirical interpretations of the canonical variates and their relationships, we may proceed a step further in our analysis of accessibility and urban growth. We may inquire into the specific way in which the linear combinations of the sets of variates have been related over time. There appear to be two distinct possibilities suggested in the literature on economic development.

One alternative is a balanced relationship over time. This involves a close interplay between changes in nodal accessibility and urban growth. For instance, as capital is invested in transportation improvements that alter network accessibility, there occurs an almost immediate response in the form of an expansion of investment in industry or an increase in the concentration of population in urban centres. Conversely, an increase in the investment in manufacturing generates a demand for additional investment in transportation and leads to improvements in accessibility. Neither nodal accessibility nor urban growth changes without triggering a concomitant reaction by the other.

A second alternative is an unbalanced relationship over time. This means the existence of a time variation involving either a 'lead' or 'lag' association between changes in nodal accessibility and urban growth. The provision of improved highway facilities might precede by some period of time an expansion of investment in manufacturing or a rapid increase in urban population. Conversely, the provision of transport investments might lag behind the demands for increased accessibility that are generated by increase in industrial or population growth in the urban centres. Both sequences provide incentives and pressures either to take advantage of the lower transfer costs provided by improved accessibility by expanding production, or to reduce certain costs of production by providing improved transport facilities.

Employing canonical analysis we may consider whether the relationships between nodal accessibility and urban growth have been balanced or unbalanced in the São Paulo study area. Because the time period for which data are available is limited to twenty years, our conclusions must be tentative. Nevertheless, the time period is sufficient to arrange the sets of variates $X^{(1)}$ and $X^{(2)}$ into four combinations:

(1) Increases in both sets of variates for the period 1940 to 1950.
(2) Increases in both sets of variates for the period 1950 to 1960.
(3) Increases in set $X^{(1)}$ for the period 1940 to 1950 and increases in set $X^{(2)}$ for the period 1950 to 1960.
(4) Increases in set $X^{(2)}$ for the period 1940 to 1950 and increases in set $X^{(1)}$ for the period 1950 and 1960.

If the relationships between nodal accessibility and urban growth have been balanced over time, the highest canonical correlations will be associated with combinations (1) and (2). If the relationships have been unbalanced, the highest correlations will be associated with either combination (3) or (4).

A tentative answer as to whether the pattern of development has been balanced or unbalanced is provided by the results of the canonical analysis listed in Table III. It is evident that the highest correlations are associated with changes in modal accessibility between 1940 and 1950 and changes in urban growth between 1950 and 1960. Apparently the relationship is unbalanced, with transportation playing a 'leading' role in the developmentary process. The coefficients of the canonical variates indicate that the highest weights must be placed on: (1) increases in accessibility to the São Paulo Maior Region between 1940 and 1950 and increases in the gross value of manufacturing between 1950 and 1960, and (2) increases in accessibility to the Pioneer Region between 1940 and 1950 and increases in urban population between 1950 and 1960.

From Table IX it can be seen that transportation as a 'lead' factor in São Paulo's economic growth has been strongest with respect to increases in the value of industrial production. The 'lead' relationship with respect to population growth is not nearly as pronounced. Quite possibly the time lag between improvement in highway accessibility and a rise in the concentration of population in urban centres is less. It should be noted that the emergence of transportation as a 'lead' factor in São Paulo's urban growth, especially with respect to industrial development, is in keeping with the theories of economic development advanced by Rosenstein-Rodan (1962) and Hirschman (1958).

The results of the canonical analysis suggests that some basic objectives of governmental policy behind highway construction and improvement are being realised. An argument frequently advanced to justify expenditures for highway development is the importance of network accessibility to regional growth. To promote economic development, the government is committed to a reduction in transfer costs as a means of overcoming barriers that limit the geographical spread of manufacturing sectors of the economy. The postwar growth of manufacturing centres along the major highways in the State of São Paulo indicates that some of the barriers to a wider geographical development of industrial economies are being overcome, especially

Table IX. Canonical Analysis of Temporal Relationships Between Nodal Accessibility and Urban Growth

Combination of sets of variates	Root	$X^{(1)}$				$X^{(2)}$				$\sqrt{\lambda}$	χ^2	D.F.
		X_1	X_2	X_3	X_4	X_5	X_6	X_7	X_8			
1	1	4·48	−2·09	0·43	2·18	−1·33	0·27	0·24	0·13	0·34	24·4	15
	2	−1·08	4·11	−1·32	−0·51	0·73	0·45	0·82	−0·62	0·23	9·4	8
	3	0·32	0·37	0·98	−0·19	−0·85	0·08	1·12	−0·18	0·09	1·5	3
2	1	3·22	1·03	−0·29	1·83	−1·48	0·16	0·05	−0·72	0·28	15·3	15
	2	−1·78	2·15	−0·51	1·08	−1·44	−0·71	0·30	0·46	0·18	5·4	8
	3	0·82	0·53	−0·86	1·07	0·30	−0·34	0·97	−0·08	0·08	1·3	3
3	1	3·19	−1·12	0·45	2·47	1·60	−0·92	0·85	0·31	0·37	29·2	15
	2	−2·30	1·97	0·16	1·31	−1·14	0·47	0·77	−0·13	0·26	10·9	8
	3	0·98	−0·58	0·15	−0·11	−1·73	0·40	−0·40	−0·21	0·09	1·5	3
4	1	7·13	3·08	1·97	9·61	4·94	−1·42	1·02	0·98	0·65	90·8‡	15
	2	3·68	6·94	−1·12	1·31	−1·14	3·67	0·77	−0·34	0·33	18·2†	8
	3	1·24	−0·59	0·44	−0·25	−0·64	0·11	0·33	0·11	0·16	4·0	3

‡ Significant at 0·99 confidence level.
† Significant at 0·95 confidence level.

with respect to transportation. Centres in the São Paulo Maior Region which have experienced increases in network accessibility have benefited by increases in their rates of economic development. Also, they have been attracting an increasing proportion of the urban population growth. If the past two decades are any guide, it would appear that there will be an increase in the number of centres which will derive economic benefits from changes in their network accessibility. This possibility is encouraging to governmental planners who believe it is essential to the continual expansion and growth of the Paulista economy.

REFERENCES

ANDERSON, T. W. (1958). *Introduction to Multivariate Statistical Analysis*, New York.

HIRSCHMAN, A. O. (1958). *The Strategy of Economic Development*, New Haven, Conn.

PIRES FERREIRA, J. (1940). Tratado de Mecanica Economica-Transportes, *Revista do Clube de Engenharia do Rio de Janeiro* (March), pp. 51–67.

ROSENSTEIN-RODAN, P. N. (1962). Notes on the theory of the 'Big Push', (ed.) H. Ellis, *Economic Development for Latin America*, London.

11 Transportation and Urban Development in West Africa: A Review

SHALOM REICHMAN†

THE introduction of modern transportation technology in West Africa between the 1880s and the present has been closely associated with political, social and economic transformation which occurred in that region during the past ninety years. Among the more spatially evident phenomena in the evolution of the socio-economic environment was the rise of urban growth poles in virtually every West African country. In this paper, a number of interactions between the transportation system and urban development will be discussed.

Three aspects deserve particular attention: the relationship between the localisation of transportation facilities and the spatial elements of the urbanisation process; the role of specific transportation factors in the evolution of urban form and activities; and, finally, the interdependence between transportation networks and flows and the development of urban systems. It is believed that these aspects reflect, *grosso modo*, both the sequence of urban development in West Africa and the more significant relationships between transportation and urban development.

Neither of the two main, available sources of information, transportation studies and urban studies, provides a full treatment of the subject. In the transportation field, the majority of studies are confined to specific modes‡ (Harrison Church, 1949; Rey, 1961; Reichman, 1965. Many works have been published on ports. The more recent are Hilling, 1969; Hoyle and Hilling, 1970), or to particular countries (Kayser and Tricart, 1957; Hawkins, 1958; Walker, 1959; Gould, 1960; Stanley (1970); only a limited number of works have a broader scope (Harrison Church, 1956; Hance, 1958; United Nations, 1962*a*; Morgan and Pugh, 1969*a*). Similarly, urban studies

† This is a revised version of a paper in 'The role of Transportation in the urban development of West Africa', presented at an international symposium on urban growth in Africa, Bordeaux, October 1970.

‡ Industries et Travaux d'Outre-Mer, *Les chemins de fer africains*, July 1963, and *Les chemins de fer africains et malgaches*, October 1969.

usually focus on the disaggregate level of the single country or the individual town (Grove and Huszar, 1964; Mabogunje, 1965, 1968; Villien-Rossi, 1966; Karmon, 1967; Seck, 1968, 1970*a*; Cotten, 1968; McNulty, 1969), though a relatively larger number of aggregate approaches have been formulated (Steel, 1956, 1961; Harrison Church, 1959; Kamian, 1963; Thomas, 1965; Vennetier, 1969; Hance, 1970*a*). In addition, much information on urban development can be found in related fields of the social sciences, in particular, sociology (Mercier, 1964; Kuper, 1965; Little, 1965; Miner, 1967; Jenkins, 1967). Contrarily, the more recent data on transportation development is often less accessible, since it was frequently compiled and analysed in detail by private consultants, with the concomitant problem of disclosure through scientific channels. This trend is closely associated with the utilisation of technically qualified foreign consultants by the newly independent states. For some of the earlier examples, see: Stanford Research Institute, *The economic co-ordination of transport development in Nigeria*, February 1961; and République du Sénégal: *Aspects des Transports Routiers au Sénégal*, Rapport BCEOM, 1962. Since then, many more consulting firms have been involved in transportation collection and analysis.

TRANSPORTATION AND URBANISATION IN WEST AFRICA

Urbanisation in West Africa is generally recognised as a predominantly colonial process (Kamian, 1963). One hundred years ago, all the capitals of the present thirteen independent states in the region either did not exist or were minor centres (Fig. 11.1). The region includes the following countries: Senegal, Gambia, Guinea, Sierra Leone, Liberia, Ivory Coast, Ghana, Togo, Dahomey, Nigeria, Niger, Upper Volta and Mali. Mauritania and Portuguese Guinea were not treated within this work. However, it has been suggested that by now one spatial element of the urbanisation process, namely the broad lines of distribution of urban centres, can be regarded as virtually complete (Hance, 1964). There is no reason not to accept this conclusion as a realistic short-term appraisal of the situation. In the long run, however, some uncertainty remains due to the possibility that boundary shifts may result in the break-up of existing states or provinces or, alternatively, in the unification of neighbouring states.

Assuming that the localisation process is all but finished, and limiting the investigation to present-day capitals, how then are transportation factors specifically related to the location of these capitals? The evolution of some of the relevant parameters is illustrated in Fig. 11.1. In terms of transportation expansion, railway building started in 1885 with the Dakar–St Louis line and reached its peak between 1900 and 1930, when it levelled off, although growth has continued until today. Road construction started at the end of the First World War, and today road mileage considerably exceeds that of railroads. The growth of capital cities is even more impressive: in absolute numbers, the aggregate population of the thirteen capitals increased from 100,000 to 1,000,000 between 1900 and 1950, whereas

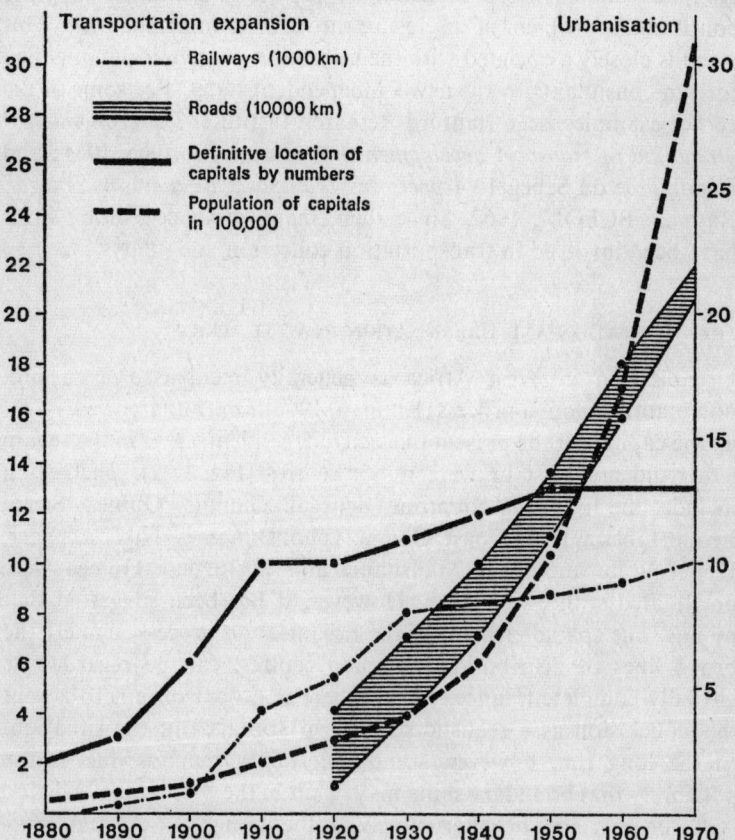

Fig. 11.1 *The evolution of selected transportation and urbanisation phenomena in West Africa, 1880–1970*

it jumped to 3,200,000 in the last two decades. This corresponds to an aggregate annual growth rate of 4 to 6 per cent, mainly since 1940. Three metropolitan areas have acquired populations between 500,000 and 1,000,000: Accra, Dakar and Abidjan, while Greater Lagos has more than 1,000,000 inhabitants. In these metropolitan areas, annual rates of increase are twice as high as the aggregate.

The definitive location of colonial capitals in West Africa was an ongoing process until the late 1940s, partly as a result of boundary shifts and partly due to the relocation of capitals within fixed territories. For instance, Accra and Lagos can be considered the definitive administrative centres of the Gold Coast and Nigeria respectively only after Ashanti was annexed to the former (1901) and Lagos Colony was merged with the protectorate of Southern Nigeria (1906). Similarly, Ouagadougou became the capital of Upper Volta only in 1948, when the number of landlocked territories in the French-controlled interior finally reverted to three. In the Ivory Coast, Abidjan was proclaimed the capital as late as 1934, preceded by Grand Bassam and Bingerville; Niamey superseded Zinder as Niger's capital in 1926.

It would be tempting to assume simply from the visual relationships in Fig. 11.1 that the location of capitals in West Africa is related to railroad expansion, while their increase in size is associated with road-building in the region. However, the causal links between transportation and urbanisation phenomena were more complex and require a careful definition of their environmental and historical context.

Contextual factors affected both transportation and urbanisation locations by limiting the number of railheads and administrative centres to one per territory. The only exceptions occurred when at least one of the conditions in Fig. 11.2 was different: the Sekondi–Tarkwa line in the Gold Coast and the Port Harcourt–Enugu section in Nigeria were built for exploitation of known mineral resources. Subsequently, a number of mineral railroads were built in various parts of West Africa without much effect on urban location (Marampa–Pepel in Sierra Leone, all railroads in Liberia, and the Fria–Conakry line in Guinea). Similarly, the requirements for transportation and urban locations were generally identical: proximity to Europe, which can be interpreted as a seaside location; centrality, either in terms of the territory as a whole or, alternatively, a central location along the coast; and, finally, accessibility to the rest of the territory, primarily to those regions with existing or latent economic potential and preferably by natural routes.

Only the constraints confronting transportation and urban locations evinced significant differences. The most formidable constraint was, of course, the physical environment and its various elements,

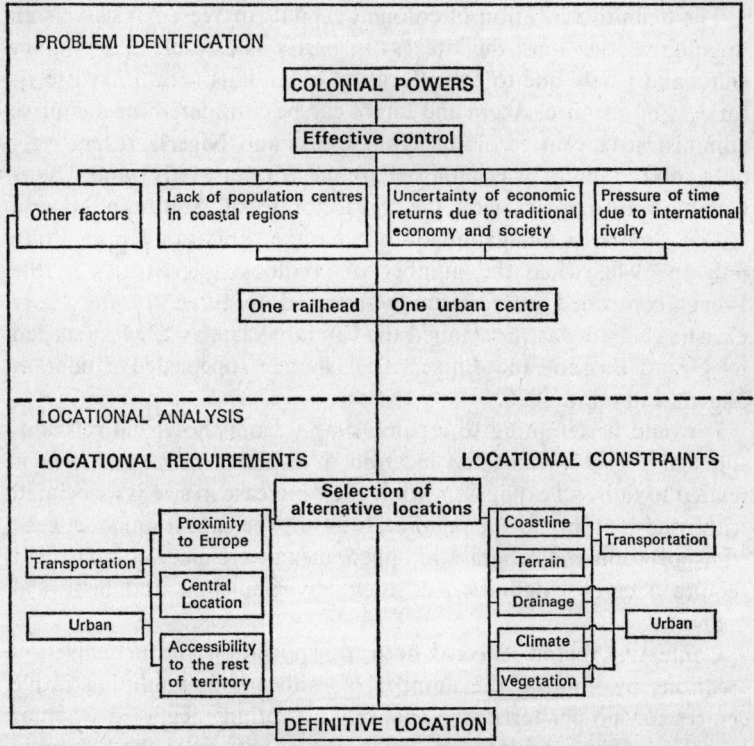

Fig. 11.2   A generalised diagram of the relationships between urban and transportation locations in West African territories around 1970

as shown in Fig. 11.2. In general, transportation location constraints were more severe: the number of easy installations for land/sea transshipments was limited, as well as the number of natural harbours. The construction of permanent ways and bridges required a greater consideration of terrain, drainage and vegetation than did urban build-up, due to both the characteristics and the greater scale of locational decisions in the transportation system. An important constraint which particularly affected urban location was the availability of climatic conditions favourable to Europeans. Notable examples are Dakar in the dry season, and the hill stations near Freetown and Accra (Dickson, 1969).

On the basis of the relationships specified so far, it is suggested that the location of railheads and administrative capitals in West Africa was affected simultaneously by contextual factors and by the requirements of transportation and urban location.† However, the importance of cost constraints in transportation was generally a greater factor in determining location. Once the convergence in space of transportation and administrative function was achieved, the well-known process of agglomeration economies took place; under colonial conditions, this process is usually associated with unbalanced growth (Hirschman, 1958; Gellar, 1967; Gauthier, 1969). Thus, every capital developed as a major rail, road and eventually air centre and, in those countries having a maritime outlet, as a seaport (Hance, 1970b). This process is not yet complete. For instance, in Dahomey, coastal Cotonou, which is by far the greater centre in size and diversity of functions, is not yet officially the administrative capital, although many government departments have moved there from Porto-Novo. Niamey still lacks a through rail connection to the Gulf of Guinea, despite the forthcoming mineral exploitation at Arlit, though it stands to benefit from the regularisation of the Niger flow after the completion of the Kainji dam. Another important aspect of the relationship between transportation and urban development is the large share of the transportation sector in the economic base of West African capitals. In Dakar, for example, transportation enterprises in the mid-1960s accounted for more than a third of total private enterprises, and the number of employees was about one-quarter of total employees in the private sector. This proportion is even greater when the public sector is added (Seck, 1970b).

The relationship between road expansion and the size increase of the capital cities in West Africa appears to be less direct than the localisation process of railheads and capitals discussed so far. It is conditioned by two major processes: first, road impact on economic growth, and, second, motor transportation as a prime instrument of social change. If the removal of the high transfer cost barrier results in economic growth, then the main urban centre will necessarily also grow, since it is usually a trade outlet or the prime region of industrial

† This relationship is clearly stated in *Rapport d'ensemble sur la situation générale du Gouvernement de la Guinée Française, 1899*, p. 116.

and service location. New jobs are offered, and the attraction of the city increases. For these opportunities to be met, labour force migrations from the countryside to the capital must occur. This is usually among the main outcomes of the social effects of transportation and particularly of motor transportation. The flexible operations of the motor vehicle and the initiative of the African driver have combined to make roads an important element in the modernisation process (Sesay, 1966; Wilson *et al.*, 1966). It should be noted, that both effects on urbanisation are gradual and cumulative, for they depend upon an ever-increasing network and require lengthy periods of time for social adjustments to develop.

TRANSPORTATION—URBAN FORM AND URBAN ACTIVITIES
IN WEST AFRICA

The importance of transportation in shaping the urban form of West African capitals is reflected in the location of transportation facilities within these cities. The basic urban form, which is found whenever town and railroad developed simultaneously, consists of a rail terminal adjacent to one side of the port, with commercial zones on the other side. Dakar, Freetown, Conakry and Lomé are good examples of this form which is essentially concentric. Roads and the retail functions which develop alongside them, have usually introduced sectoral elements. This layout is remarkably similar to that proposed for new Nigerian towns in 1939 (Morgan and Pugh, 1969*b*).

Another, and possibly more important, effect has been the influence that urban form exerts upon urban transportation. Here, some typically African relationships can be distinguished.

Site constraints

It has been shown above that in most cases general locational factors favoured a waterside site for West African capitals (Harrison Church, 1967). However, this inevitably resulted in the creation of serious constraints on the development of the cities' built-up areas. In almost all the capitals, physical constraints have caused significant distortions of built-up areas away from an ideal circular, rectangular or star shape. Those cities which have relatively few physical obstacles, such as Accra and Ouagadougou, have created man-made constraints. Both cities have close-in airports, and in Ouagadougou an artificial lake forms a barrier to the northward expansion of the city. A general consequence of the irregular shapes is an increase in distances, and the concomitant cost of infrastructure and operating costs of movements. In addition, water obstacles necessitate expensive bridging with strong funnelling effects, which may create traffic congestion even where vehicle flow is relatively low. Outstanding examples are Lagos and Abidjan, where a centrally located island in each city is linked to the mainland by two roughly parallel bridges.

Social area differentiation

From their inception, West African capitals have been characterised by the dual composition of their population and its reflection in the dual nature of their residential quarters. High-class housing, inhabited by Europeans and later by high-income Africans, is located on favourable sites either in the centre (Dakar-plateau, Abidjan-plateau, Accra-ridge) or, more frequently, in low-density areas in the periphery (Abidjan–Cocody; Lagos–Ikoyi and Apapa; Freetown–Wilberforce; Monrovia–Sinkor). In extreme contrast is high-density, low-income housing, often reminiscent of shanty-towns. This housing also tends to have specific locations: lowlands or the periphery of the built-up area. In coastal capitals, the large, generally unplanned low-class housing areas form a crown on the landward side of town, whereas high-class housing tends to be located on the seaward margin. The polarisation of residential areas in opposite corners of town naturally reinforces the process of urban sprawl and its concomitant increase in the friction of distance. It is not unusual to find distances of 10 km from one end to the other of built-up areas in towns with only 100,000 inhabitants. The most extreme example of a 'stretched' town resulting from physical and social factors is probably L-shaped Monrovia, with a distance of 25 km from New Kru to Sinkor. In all four of West Africa's large metropolitan areas, such distances are not very uncommon, perhaps with the exception of more compactly built Accra (Karmon, 1967). Another negative effect of transportation on urban form and activities is the poor quality and design of the transportation infrastructure, sometimes causing severe congestion with even relatively small traffic flows. In Freetown, for instance, over 80 per cent of the streets, in 1965, were not wide enough for parking, and 30 per cent were either unmotorable or single-lane. Many streets even lacked pavements, and were lined by storm water drains (Harvey and Dewdney, 1968).

The factors mentioned so far operate mainly on the supply side, as cost elements of a transportation production schedule. As a tentative conclusion, it is suggested that the provision of the transportation infrastructure in West African capitals is relatively costly, due to high capital outlays per mile of road and also to the marginal addition of road mileage *per capita*. Transportation demand and its corollary, travel behaviour, are affected mainly by the interaction among socio-economic characteristics of the population, land-use patterns and transportation system characteristics. In West Africa, two factors are outstanding: large segments of the urban population have extremely low incomes; but, at the same time, motorisation rates in West African capitals approach those of many European towns. Each factor has diametrically opposed effects on the activity patterns in the city.

Income, or monetary budget constraints, is by far the most important factor in African travel demand in general and urban travel in particular. Time budgets, or considerations of time saving, play a minor role in traffic generation and modal choice. Household surveys suggest that average transportation and communication expenditures in urban families do not exceed 7 per cent of total consumption (Engelbrecht, 1965; Birmingham *et al.*, 1966). As an illustration, average expenditure *per capita* on transportation in Accra has been estimated at 338 shillings per annum (in the early 1960s). A similar estimation of the average expenditure on public short-distance urban transportation was made in Ibadan (in 1964) and totalled £4 per annum. It would be wrong to assume that these mean values represent normal distributions. In reality, a large section of the urban population cannot afford to pay for transportation. Recently it was found that over 50 per cent of the African workers in Abidjan, which is in aggregate probably one of the wealthiest cities in West Africa, walk to work (Demur, 1969). Travel behaviour of petty traders from the Cap Vert region to Dakar also indicates the paramount importance of even small cash outlays for the use of public transportation (Seck, 1967). In summary, the effect of income is comprised of two different elements: the general problem of low participation by the African urban population in the wage-earning labour force results in a scarcity of cash which affects travel behaviour. A different factor altogether is the low income of significant numbers of wage-earners. In this case, transportation expenditures are necessarily minimal and pay mainly for travel to and from work. According to officials of the bus company in Abidjan, fluctuations in passenger demand during a given month are caused by the availability of cash as a function of pay-days.† A tentative conclusion can be put forward that large sections of the urban population in West African capitals still cannot fully take part in the activity system due to their income constraint.

Motorisation rates in West African capitals are not easy to determine: information is generally not available on the number of registered vehicles which are no longer in operation; little is known of vehicles which are registered in one place but operate in another; and, furthermore, deficiencies in population estimates undermine the accuracy of information on the degree of motorisation. Given the above reservations, motorisation rates in West African capitals can be estimated as ranging from 40 to 100 vehicles per 1,000 inhabitants. As a rule of thumb, it is suggested that 60 to 70 per cent of the private cars and up to 40 per cent of the commercial vehicles in every country operate in the capital city, with the exception of Nigeria. In recent years, it seems that in most countries of the region the increase in motor vehicle parks has exceeded the growth of population, so that in the short run the motorisation rate is likely to increase.

The chief reason for this high rate is clearly connected to the travel habits of the Europeans in the population, and probably also to the travel habits of the upper levels among African civil servants. This segment of the population demonstrates a typical pattern of car-captive behaviour: virtually all European households have a private car available and in many cases operate two cars. Assuming an average family size of four, the rate of motorisation ranges from 250 to 375 cars per 1,000 inhabitants, which is close to the saturation point. Unlike the average African household, income constraints have relatively little effect on car acquisition, not only because of the generally high income of Europeans, but also due to the various schemes of cost participation by employers which amount to a partial subsidisation of private vehicles. Another cause for this high car ownership and usage is the spatial distribution of European residential areas: they are usually far out and tend to have low housing densities, so that alternative means of transport are inefficient. However, psychological factors should also be examined for a possible explanation, for even in those areas where public transportation is

† Information supplied SOTRA to C. Gabriel, graduate student at the Institut de Géographie Tropicale, Université d'Abidjan.

provided, almost no use is made of it by Europeans, except for taxi services. Possibly the same social segregation which is evidenced in housing preferences is extended to the choice of the most private transportation modes.

Extensive motorisation is West African capitals permits, in fact, the introduction of a highly specialised activity system into the much less diversified activities of the average African: thresholds of goods and services are lowered, and their spatial agglomerations are reinforced. Perhaps more than any other single factor, the private car has been a means of superimposing a 'modern' and often deceptive veneer upon the African urban scene. Last, but not least, the large traffic flows generated by private cars have had a dominant influence on the accessibility pattern of the cities, since they have dictated to a large extent which links in the existing road network need improvement.

An important evolution in the urban transportation system in the last decade has been the creation of publicly-owned bus services. The market penetration of the fixed-route, unitary tariff bus service has been contingent upon a number of factors: the suppression of its closest competitor, the collective taxi or mini-bus; the size of the population, the number of wage-earners, and the density of potential user population along the routes. It has already been mentioned that the car-captive population is not likely to use buses, and the same is probably true of bicycle owners in the smaller cities and moped-riders in the larger. It has been estimated that the minimum size of towns for public bus service to be profitable is 100,000 inhabitants (Podoski, 1965), but it is clear that the four largest West African metropolitan areas are likely to benefit first from bus transportation. Abidjan shows probably the greatest bus penetration with 73,000,000 journeys in 1969, as against 15,000,000 in 1963. The unitary tariff is 20 fr CFA.† With an average of 145 bus-trips *per capita* last year, Abidjan had a bus-user rate at least twice as high as Bamako, Freetown and Cotonou. Incidentally, the 'big four'—Lagos, Accra, Dakar and Abidjan—also have substantial suburban rail traffic.

It remains to be seen what impact bus service will have on the urban form and on the urban activity system in West African capitals. On the one hand, buses were introduced at a late stage in the development process, and they have had little effect on either the road network configuration or on the spatial distribution of built-up areas and land-use patterns. On the other hand, it is quite likely that regular

† Information supplied by SOTRA.

public transportation will increasingly become the principal mode of transportation for the mass of the urban population, and hence a vital tool for the participation of Africans in the activity system. There is no doubt that a serious research effort is required to determine the adaption of an urban form which was partly shaped by private cars to an activity system generally based on newly introduced, regular public transportation.

TRANSPORTATION AND URBAN SYSTEMS IN WEST AFRICA

Urban systems are an inseparable part of the functional organisation of space. Three levels are usually investigated: (1) the urban field which includes urban–rural interactions; (2) a system of central places within a given area, generally comprised of the whole national territory; (3) the supra-national level, where metropolitan areas acquire specific functional dominance over urban systems in other, not necessarily contiguous regions. It is suggested that at each of these levels, particularly elements of the transportation system have contributed to the organisation of space.

At the lowest level, the main contribution of transportation is that of cost reduction or, put alternatively, of allowing a more efficient utilisation of resources. Numerous examples are cited about the dramatic reduction in transfer costs following the introduction of railways and roads in West Africa: in one case it has been estimated that road transport charges would have been one-twentieth of human porterage costs (United Nations, 1962*b*). Given this cost ratio, and assuming an initial 'isolated state', the reduction of transfer costs following the construction of one road originating from the market centre would greatly increase the margin of cultivated areas. Actually, if the usual Von Thunen assumptions are held, the area whose product could be sold at the market would be nearly seven times the size of the original circle (Walters, 1968). The savings of transfer costs might eventually be much greater, once the economies of scale inherent in modern transportation begin to be felt.

The effect of the removal of the transportation cost barrier is not limited to urban–rural relationships. However, at the higher level, another transportation factor assumes greater importance, namely spatial differentiation due to network location. A number of studies have established the existence of a hierarchy of central places in West African countries or in smaller area units within one country (Grove and Huszar, 1964; Mabogunje, 1965, 1968; Cotten, 1968; Abiodun, 1968. In most cases, transportation functions were either implicit or explicit in the rank evaluation, but little was done to isolate the network effect. A simple way to do this is to measure the local degree of the network, that is, the number of links connecting each central place with others. A 1967 rough estimate on the frequency distribution of the local degree of Nigerian administrative centres suggests that 60 per cent of the centres had two or less all-weather links with other nodes, and 17 per cent four or more links.† A significant level of local degree would be six linkages, which corresponds to the more advanced stage of economic integration of a Löschian landscape, where larger centres are directly linked, by-passing smaller centres (Haggett, 1965). In Nigeria only three centres have six linkages, as against two centres in Upper Volta‡. A temporary conclusion, which can be drawn from the relative scarcity of high levels of local degrees of linkages, is that networks' construction and maintenance costs still greatly exceed the volume of traffic. This is yet further evidence of the indivisibility of production factors inherent in transportation networks.

Another network effect on urban systems is the connectivity of the network as a whole. Initially, the increasing connectivity of transportation networks in West Africa was discussed as an evolutionary process (Taaffe *et al.*, 1963; Reichman, 1969). Subsequently, more rigorous parameters of network topology were defined, in particular with respect to the development of the road network in Ghana (Yeates, 1968). The simplest tool used was the Beta-index, which is the quotient of the number of links (arcs) in the network, divided by the number of nodes (vectors). It was found that the network grew from the minimum of $0·50$ to a simple connected network (Beta $= 1·00$), then to an ever-increasing degree of connectivity. The general case, however, seems to be more complex: there is no doubt that the Beta-index increases at early stages of development; but after a certain point the marginal increase becomes negative, so that eventually Beta-indices in more developed countries are negatively correlated with several indicators of economic development (Kansky, 1963). The reason for this evolution may again be found in the lumpy nature of transportation investments, combined with the 'lead' role of transportation in relation to regional growth (Gauthier, 1968). In other

---

† *Nigeria, Map of Internal Communications*, 1:3,000,000, Lagos 1967.
‡ *République de Haute Volta: Carte routière* 1:1,000,000, Dakar 1967.

words, once the basic network is connected, the number of centres growing on the existing network exceeds the addition of new links to nodes outside. It would be interesting to investigate the relationships between changes in the Beta-index and the development of the urban system in general, and the rise of major growth poles in particular.

Both the cost-saving effect and the network effect discussed above require a number of basic conditions before their influence can be felt. These conditions were summed-up fifteen years ago, and they do not seem to have changed significantly (Capet, 1958):

(1) Complimentarity between production potential and transportation, commercial or marketing system.
(2) Sufficient public funds to maintain the network during its initial excess-capacity stage.
(3) Propensity of individuals to accept the changes wrought by the introduction of modern transportation.

A final illustration of the relationship between transportation and urban systems is the emergence of supra-national centres in West Africa. Unlike the predominance of Dakar during the colonial period, which was essentially imposed from above, the new pattern of dominance is related to the relative location of large metropolitan centres within the region. The main transportation element involved in this evolution is the presence of regional air routes in West Africa. Each of the four metropolitan centres in the region has developed an air hinterland, which can be defined by traffic flows or flight frequencies on the one hand, and relative flight rates on the other.

The most extensive hinterland is that of Abidjan, whether measured in terms of relative costs or higher relative frequencies (Fig. 11.3). Abidjan derives its increasing importance from a variety of factors: first, its central location which makes it an ideal transfer-point from long-distance inter-continental services to medium-distance regional flights. Second, strong political and economic ties among the 'entente' countries (Ivory Coast, Upper Volta, Niger and Dahomey) naturally resulted in increasing links between these states and their main economic and political centre, Abidjan. Third, air transport networks have strong 'funnelling' effects as a result of their high-cost equipment. Hence, a typical domestic air network converges upon the capital city but is not directly linked with foreign centres. Furthermore, due to demand limitations, not all capitals are linked by direct

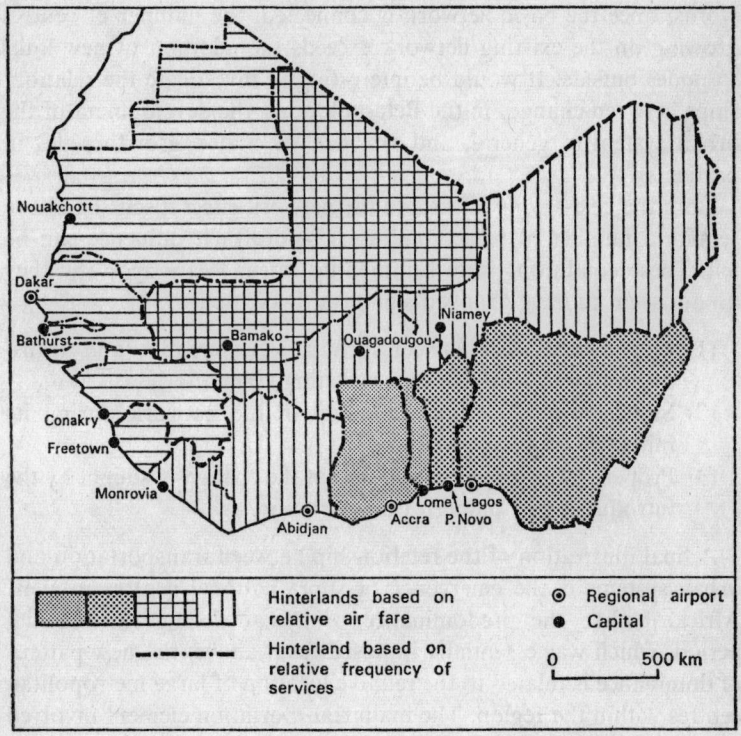

Fig. 11.3   Airport hinterlands in West Africa, 1970

routes, so that a modified network-effect is achieved, a network which tends to reinforce those air centres which already have a relatively high frequency of services to other centres in the region.

The net result of the emergence of the supra-national centres should not be underestimated, for it allows the introduction of highly specialised functions, whose threshold would normally exceed the population or purchasing power of most individual West African states. Thus, Dakar may in fact represent for certain purp oses a population of almost ten million inhabitants, the same as Accra, whereas Abidjan serves a hinterland with fifteen million inhabitants. Lagos easily has a hinterland twice or even three times as populous. The less advantageously located capitals, such as Niamey and Freetown, are definitely affected by their relative isolation from the fastest-growing urban centres in West Africa.

REFERENCES

ABIODUN, J. A. (1968). Central place study in Abeokuta Province, southwestern Nigeria, *J. Reg. Sc.*, **8**(1), 57–76.

BIRMINGHAM, W. *et al.* (1966). *A Study of Contemporary Ghana*, vol. 1. The Economy of Ghana, London, p. 111.

CAPET, M. (1958). *Traité d'économie tropicale. Les économies d'A.O.F.*, Paris, pp. 139–41.

COTTEN, A. M. (1968). 'Les villes de Côte d'Ivoire', *Bull. Ass. Geogr. Français*, pp. 223–38.

DEMUR, C. (1969). *Les Transports Urbains à Abidjan*. Mémoire de Maîtrise, Université de Paris, Institut de Géographie.

DICKSON, K. B. (1969). *A Historical Geography of Ghana*, Cambridge, pp. 254–61.

ENGELBRECHT, P. (1965). Ibadan City Council Bus Service, Western Nigeria, *Union Internationale des Transports Publics (UITP)*, *Revue*, **14**, 105–11.

GAUTHIER, H. L. (1968). Transportation and the growth of the São Paulo economy, *J. Reg. Sc.*, **8**(1), 77–94. See pages 167–189 of this volume.

GAUTHIER, H. L. (1969). Transportation and regional economic development, Paper presented at the USA–Israel seminar in geography, Jerusalem (July).

GELLAR, S. (1967). West African capital cities as motors for development, *Civilisation*, **17**, 254–62.

GOULD, P. R. (1960). *The Development of the Transportation Pattern in Ghana*, Evanston.

GROVE, D. and HUSZAR, L. (1964). *The Towns of Ghana, the Role of Service Centers in Regional Planning*, Accra.

HAGGETT, P. (1965). *Locational Analysis in Human Geography*, London, p. 82.

HANCE, W. A. (1958). Transport in tropical Africa, in *African Economic Development*, pp. 85–134.

HANCE, W. A. (1964). *The Geography of Modern Africa*, New York, p. 57.

HANCE. W. A. (1970a). *Population, Migration and Urbanization in Africa*, New York.

HANCE, W. A. (1970b). Ibid., p. 234.

HARRISON CHURCH, R. J. (1949). The evolution of railways in French and British West Africa, *Comptes rendus du XVIe Congrès International de Géographie*, Lisbon, Tome IV, pp. 95–114.

HARRISON CHURCH, R. J. (1956). The pattern of transport in British West Africa, in *Geographical Essays on British Tropical Lands*, London, pp. 53–76.

HARRISON CHURCH, R. J. (1959). West African urbanization, a geographical view, *The Sociological Review*, **7**(1), 15–28.

HARRISON CHURCH, R. J. (1967). Urban problems and economic development in West Africa, *J. modern Afr. stud.*, **5**, 511–20.

HARVEY, M. and DEWDNEY, J. C. (1968), Planning problems in Freetown, (eds.) C. Fyfe and E. Jones, *Freetown, a Symposium*, Freetown, pp. 177–95.

HAWKINS, E. J. (1958). *Road Transport in Nigeria*, Oxford.

HILLING, D. (1969). The evolution of the major ports of West Africa, *Geog. J.* (September), 365–78.

HIRSCHMAN, A. O. (1958). *The Strategy of Economic Development*, New Haven, Conn., chap. 4, chap. 10.

HOYLE, B. S. and HILLING, D. (eds.) (1970). *Seaports and Development in Tropical Africa*, London.

JENKINS, G. (1967). Africa as it urbanizes, an overview of current research, *Urban Affairs Quarterly*, **2**(3), 66–80 (March).

KAMIAN, B. (1963). Les villes dans les nouveaux états de l'Afrique occidentale, *Tiers Monde*, pp. 65–80.

KANSKY, K. J. (1963). *Structure of Transportation Networks*, University of Chicago, Department of Geography Research Paper 84, p. 61.

KARMON, Y. (1967). Accra-Tema, Das Werden einer afrikanischen Grosstadt, *Erdkunde*, **21**(1), 33–48.

KAYSER, B. and TRICART, J. (1957). Rail et Route au Sénégal, *Ann. de Géogr.*, **66**, 328–50.

KUPER, H. (ed.) (1965). *Urbanization and Migration in West Africa*, Berkeley.

LITTLE, K. (1965). *West African Urbanization, a Study of Voluntary Association in Social Change*, Cambridge.

MABOGUNJE, A. L. (1965). Urbanization in Nigeria, a constraint on economic development, *Economic Development and Cultural Change*, **13**, 413–38.

MABOGUNJE, A. L. (1968). *Urbanization in Nigeria*, London.

MCNULTY, M. L. (1969). Urban structure and development, the urban system of Ghana, *The Journal of Developing Areas*, **3**, 159–76.

MERCIER, P. (1964). L'Urbanisation au Sénégal, (ed.) W. Frohlich, *Afrika im Wandel seiner Gesellschaftsformen*, Leiden, pp. 48–70.

MINER, H. (ed.) (1967). *The City in Modern Africa*, New York.

MORGAN, W. B. and PUGH, J. C. (1969a). *West Africa*, London, pp. 581–622.

MORGAN, W. B. and PUGH, J. C. (1969b). Ibid., p. 460.

PODOSKI, J. (1965). Organization of public transport in the developing countries, *UITP International Congress*, p. 8.

REICHMAN, S. (1965). *Air Transport in West Africa*, Institut de Transport Aérien, Paris.

REICHMAN, S. (1969). Structure des réseaux aériens en Afrique occidentale, *Information Géographique*, **33**(2), 61–71.

REY, J. P. (1961). *L'aviation en Afrique Noire d'expression française*, Paris.

SECK, A. (1967). Communication at the Seminar on Regionalism in West Africa, University of Dakar (December).

SECK, A. (1968). *Les Grandes Villes d'Afrique et de Madagascar, Dakar, La Documentation Française*, pp. 3505–6.

SECK, A. (1970a). Dakar, métropole ouest-africaine, *IFAN*, Dakar.

SECK, A. (1970b). Ibid., p. 53.

SESAY, S. M. (1966). Drivers in the transport industry: a case study of road transport in Sierra Leone, *Sierra Leone Studies*, **19**, 86–97 (July).

STANLEY, W. R. (1970). Transport expansion in Liberia, *Geogr. Rev.*, **60**, 529–47. See pages 87–103 of this volume.

STEEL, R. W. (1956). The towns of West Africa, in *Geographical Essays on British Tropical Lands*, London, pp. 39–48.

STEELE, R. W. (1961). Towns of tropical Africa, in *Essays on African Population*, London, pp. 249–78.

TAAFFE, E. J. *et al.* (1963). Transport expansion in under-developed countries, a comparative analysis, *Geogr. Rev.*, **53**, 503–29. See pages 32–49 of this volume.

THOMAS, B. H. (1965). The location and nature of West African cities, (ed.) H. Kuper, *Urbanization and Migration in West Africa*, Berkeley, pp. 23–38.

UNITED NATIONS (1962a). *Transport Problems in Relation to Economic Development in West Africa*, E/CN 14/63 (April).

UNITED NATIONS (1962b). Ibid., p. 12.

VENNETIER, P. (1969). Le développement urbain en Afrique Tropicale, *Cahiers d'Outre Mer*, **85**, 5–62 (January–March).

VILLIEN-ROSSI, M. L. (1966). Bamako, capitale du Mali, *Bul. IFAN*, B 1–2, 249–380.

WALKER, G. J. (1959). *Traffic and Transport in Nigeria*, Colonial Research Studies 27, London.

WALTERS, A. A. (1968). *The Economics of Road Users Charges*, IRBC Report EC-158, chap. 5.

WILSON, G. W. *et al.* (1966). *The Impact of Highway Investment on Development* Washington, pp. 192–211.

YEATES, M. H. (1968). *An Introduction to Quantitative Analysis in Economic Geography*, New York, pp. 117–20.

12 Towards a Theory of Transport and Development

GEORGE W. WILSON†

THE variation of results in studies of transport and development may be explained by differences in two main factors: (1) the creation of economic opportunity and (2) the response to economic opportunity. The first depends upon the quality and quantity of resources in the regions served, the actual change in transport rates and service, and commodity price levels. The second depends upon an awareness of opportunity and what may be broadly defined as attitudes toward economic change.

ECONOMIC OPPORTUNITY

The extent of the economic opportunity created has two main dimensions, both of which are functions of natural resources and the rate and service changes in the transport sector. Human resources are ignored at this point and will be discussed later.

The resource base

In every case where a road opened up new territory, the soil or forest conditions in the area determined not only the type of activity but also much of the increase in output. The suitability of the soils when coupled with feeder roads is mainly responsible for the sharp increases in production that took place. Profit prospects for agriculture and forestry widened appreciably in most of the cases because of a combination of higher yields in the areas, rising prices, and declining freight rates. Even where prices of the major products were declining or stable over the period under review, the increase in yields, combined with reduced freight charges and improved service, increased growers' net receipts. For the most part, prices were determined in markets outside the local area, which meant that rising local supplies exerted little or no influence on price. The general picture thus fits a

† This is a shortened version of Chapter 8 of Professor Wilson's *The Impact of Highway Investment on Development*, published in 1966.

model of perfect competition, with the demand schedule perfectly elastic at the externally determined, though variable, price. The stimulus to production was due to declining unit costs of production and distribution. The stimulus was accentuated in some cases by rising prices or partially offset by declining prices. Where prices were falling and yields not increasing, the key to development was the decline in transportation costs. In a situation with improving yields or rising prices, reduced freight rates merely provided a further stimulus, and the new transport capacity was permissive and responsive rather than causal.

The transport sector

The extent of apparent economic opportunity in both the transport and nontransport sectors is a direct function of the natural resources that are made more accessible. There may be divergent effects between the sectors, due to rate and quality changes. The extent of the negative impact on transport of lower rates depends upon the elasticity of demand for transport and the behaviour of unit costs with changes in volume. It is probably fair to say that the total amount of economic opportunity created varies inversely with rate levels, since the stimulus to increase production probably more than compensates for any possible increase in unit costs of providing transport service. If unit costs decline with volume, and assuming appropriately high demand elasticities, lower rates may benefit both producers of commodities and suppliers of transport. The immediate result is the creation of an excess capacity in vehicles which probably depresses rates below levels that would be sustainable over time.

The excess capacity may have a lasting advantage, however. First, it provides some experience in business enterprise that, while discouraging to some, might prove salutary to others. Second, sharp stimulus to production and new settlement is due in most cases to other factors as well as to unduly low rates. The gains appear to be sustainable even in the face of later rate increases. It is even possible that lower net income per unit of agricultural output, once production is established, induces a greater output to maintain the producer's overall income, although this is by no means certain. Better still, it might encourage the use of fertilisers or other improvements that lead to lower unit costs. These improvements would have a much more permanent influence on yields.

RESPONSES TO ECONOMIC OPPORTUNITY

Attitudes and awareness influence the response of individuals to the creation of economic opportunity. The response may be zero, negative, or positive in terms of developmental impact and is broadly bound up with aspects of culture, social relationships, individual psychology, and levels of well-being. It is here that the economist steps gingerly into that overflowing category of 'other things' in the qualification 'other things being equal'. Formal economic models make little reference to these reactions and presume responses roughly applicable to institutions mainly relevant to western economic systems. Despite the serious misgivings of some economists,† the bulk of growth theory continues to be overly schematic, general, and aggregative. Some economists have turned completely to non-economic expanations or have raised serious questions whether economic development has much to do with economic matters at all, at least in the form of contemporary micro- and macro-analysis developed in the West. The purpose here is not to analyse the growing guilt feelings of economists in understandably de-emphasising institutional factors but to suggest that a meaningful interpretation, especially of localised growth, cannot ignore them.

In considering the question of attitudes that influence the response to economic or other types of change, the economist must tread warily. Yet tread he must if he is to derive relevant conclusions and make useful policy suggestions. Essentially the question is: under what circumstances and to what extent will economic opportunity be exploited in such a way that net output *per capita* rises? Additional transport capacity generates new opportunities for pecuniary gain. There is no apparent consistency in the extent to which such opportunities are seized or in the apparent consequences.

The main factors influencing the response to new transport capacity are: (a) awareness of its potential, (b) the availability of finance, and (c) the magnitude of the possible benefits relative to alternative investment options.

---

† For example, 'in those countries where growth seems most essential for human welfare, problems outside the conventional limits of economics are surely paramount. Indeed, a strong argument can be made that the problem of under-development will not be solved until economics has achieved a more compatible marriage than now prevails with other social sciences'. (Hoselitz *et al.*, 1962).

Awareness of the new potential

Obviously, to evoke any response people must know that new economic opportunity has developed. The extent of awareness of the consequences of additional transport capacity depends in large part on the number of people who are directly influenced. The greater the population in the area through which the new or improved facility runs, the more extensive is the knowledge of what it might accomplish. More people can sense the fact that something new has happened that may be of benefit to them. In areas where communication is defective, this is obviously important. But beyond the mere numbers of people affected, there is the question of accessibility. A pipeline traversing a heavily populated area cannot evoke much response, whereas an unlimited-access road can. The unlimited access of roads in the early stages of development of any region has an awareness effect that serves to induce a larger number of people to take advantage of new economic potential. This contrasts with the direct economic stimulus of limited-access highways where congestion or local bottlenecks slow down traffic and raise the costs of transport. Indeed, it is this feature that underlies the frequent assertions that the indirect or spillover effects of road transport are more important than the direct reductions in user costs and faster transit. For example, R. S. Millard argues that unlike developed countries, in overseas territories, 'the benefits from road construction are almost entirely in the form of new development from the traffic which the new road will generate' (Millard, 1959). The so-called 'openness of roads'† acts as a kind of advertisement for its own economic potential, to which many individuals may respond since it is not a private nor a closed public facility accessible to only a handful of owners or employees.

Railroads are in a peculiar situation in this regard. To increase accessibility requires more stations along the line and hence shorter hauls and more frequent stops, which is inconsistent with economical operation. Increased accessibility for railroads is thus purchased at the expense of higher costs which in turn require higher rates if the facility is to be self-supporting. While increased accessibility is stimulating in itself, the higher rates reduce the magnitude of economic opportunity created and thereby offset to a greater or lesser extent the stimulus to development due to improved access.

† This suggestive phrase is from Haefele, 1963*a*.

Railroads have been most successful in moving large amounts of goods over long distances from a specific region to a port or a major consuming centre. The developmental impact along the right-of-way is generally far less than for facilities that are more open.

Effects of different kinds of investment

In underdeveloped economies, there are important variations in the *kind* of impact to be expected from alternative investments in the field of transportation and elsewhere. Some types of capital formation have spillover effects, in addition to direct effects, that differ greatly from others. Indeed, it is precisely these varying possibilities that economic planners seek to exploit in attempting to maximise the rate of growth of output. But economic analysis ranks investment projects by some version of 'expected rate of return' or benefit–cost ratio, both of which typically ignore many indirect and noneconomic effects. Yet the creation of economic capacity is only permissive. Effective utilisation and augmentation require attitudes, abilities, and incentives that cannot be taken for granted in most underdeveloped economies. It has often been stressed that the key to sustainable economic growth is a change in attitudes toward work, business, thrift, and so on. Therefore, another way of ranking investment projects is by their effect on attitudes. In this respect, certain investments may provide a greater catalytic effect than others whose immediate payoff in increased output is far superior.

Thus investment outlets might usefully be construed in terms of their relative influence on attitudes *vis-à-vis* labour productivity through an increase in the capital–labour ratio. The first refers to a qualitative change in the labour input, while the latter refers to a quantitative change in the amount of capital relative to the quantity of labour. Both affect productivity, although in different ways. Changes in attitudes may positively influence productivity with no increase in physical capital, that is, by altering the duration and intensity of labour or stimulating entrepreneurial activity. Likewise, it is possible to raise efficiency by additional machinery without changing attitudes, through adaptation of the equipment to contemporary customs and attitudes. These are extremes, of course. In most instances, some degree of change in attitude will be induced by additional equipment. That is, workers and management will have to adjust somewhat to new techniques or even to an extension of existing facilities, and these in turn may partly be arranged to suit existing attitudes.

But the point is that it is possible to array alternative investment possibilities in terms of their direct impact upon efficiency through their effects on the quantity of directly productive capital on the one hand and their influence on attitudes on the other.

Investments in health, education, and propaganda, for example, are direct attacks on attitudes and abilities that do not create capital for directly productive activities. At the other extreme is investment

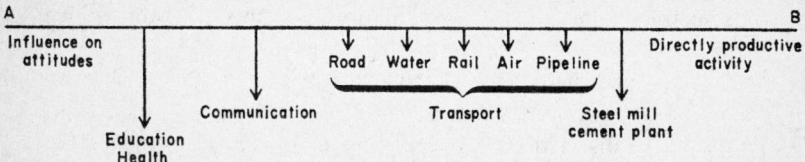

Fig. 12.1   Investment options and attitudes

in a particular factory which may have no influence on attitudes or at best influence only those directly employed. It also appears that the overall incidence of these two extremes, in terms of numbers of people affected by the investments, is significantly different. Investments in directly productive facilities usually affect a far smaller number of people than investments in health, education, or general communications.

Between these two extremes are a series of other investment options that influence attitudes and either facilitate the development of directly productive activities or are a form of such activity themselves. We may therefore place particular kinds of investment along a continuum as portrayed in Fig. 12.1. At one extreme *A* are those investments that affect only attitudes but have no impact on efficiency directly, while at the other extreme are investments that only have such an impact. As one moves from *A* to *B*, the relative importance of the direct influence on efficiency through a change in the capital–labour ratio increases *vis-à-vis* the impact on attitudes, and conversely when moving from *B* to *A*. The particular investments arrayed on the scale are based on intuitive judgment for the most part. However, the discussion will examine the position of transport investments as a whole and the position of particular forms of transportation on this continuum.

This taxonomy does not suggest that if an investment is placed closer to *B* than to *A*, it necessarily loses much of its influence on attitudes. It is only the ratio that varies as one moves along the scale. There is no attempt to apply isoquant analysis nor to suggest the substitution possibilities between investments in qualitative change in labour inputs and quantitative change in (directly productive) equipment. The nature of the relative payoffs are too intangible and amorphous to warrant application of the isoquant approach. Furthermore, the isoquant approach is concerned with an efficient allocation of capital and other resources *a priori*, whereas the interest of the present study centres on the explanation of events following a given investment.

Harvey Leibenstein has adopted an approach bearing some affinity to the above, but since his interests were on optimal allocation criteria, his use of the distinctions between the two types of investment which he calls 'human' and 'capital goods' is quite different from that in this study. Leibenstein's critique of the 'social marginal productivity' criteria is consistent with the views expressed above. However, his use of a standard involving effects on the supply of entrepreneurship, future savings habits, population growth, and so on in the form of the familiar isoquant approach implies a degree of possible quantification that seems unrealistic (Leibenstein, 1957a). In this sense, Hirschman's comment that Leibenstein's 'criticism seems likely to result in an agnostic "it all depends" attitude since it seriously impairs the usefulness of the SMP (social marginal productivity) criterion without replacing it by a manageable new instrument' seems valid (Hirschman, 1962). But Leibenstein's critique of SMP is valid enough—it is only his substitute criteria that seem defective.

Rosenberg has pointed to a classification roughly analogous to that in the text. For example, he argues that

Economic growth is, in many important respects, a learning process, a process whereby the human factor acquires new skills, aptitudes, capabilities, and aspirations. And the pattern of resource use which may maximise output from a given stock of resources may or may not generate the qualitative changes in the human agent which are most conducive to the growth of output in subsequent time periods.

Neo-classical economics fails to capture much of the explanation for the growth in productivity because of the failure to consider a variety of feedback mechanisms. We fail to consider, for example, the impact upon productivity of certain kinds of economic activities as opposed to others—such as manufacturing vs. agriculture. Different kinds of economic activities have different kinds of effects upon the productivity of the human agent . . . (Rosenberg, 1964).

The present aim is to analyse the extent to which investments influence attitudes aside from their direct technical efficiency.

Figure 12.1 implies that transport investments lie intermediate on the scale between *A* and *B*. This position requires some comment. Investments in transportation have a wider geographical dimension than almost any other. Therefore, they affect a far greater number of people and in a more intimate fashion than does a factory or other facility requiring a specific location. Furthermore, access to the latter types of investment is limited mainly to employees, although their products may be widely distributed. However, although an investment may affect large numbers of people, this does not guarantee that the impact will be favourable. Often such an impact will make certain people cling even more compulsively to traditional ways. Much depends on the magnitude and nature of the investment as well as on its geographical dispersion and the number of people directly influenced. But even wide distribution of products will not affect attitudes nor give rise to much additional economic opportunity unless the product is a producer's good, in which case it will be acquired mainly by people already 'developed' in an entrepreneurial or business sense. Even so, we cannot argue confidently that a series of small factories whose total cost equals that of one major transport facility will necessarily have a smaller favourable impact than the facility itself. The discussion here is mainly heuristic and admittedly tentative. I am indebted to Professor E. E. Hagen for these and other caveats regarding this section of the chapter.

Perhaps location in a densely populated area may increase the exposure of a new investment. However, it is in urban areas that one finds a dense population and the greatest degree of development and the widest range of economic opportunity already existing. It is in the more traditional rural areas that change is urgently needed since generally the bulk of the population is rural. Thus, a steel mill in an urban complex will not have much influence on attitudes despite the fact that large numbers of people are aware of it. Transport facilities, on the other hand, not only can be used directly by many people but when extended into rural areas can bring a greater proportion of the most traditional aspects of a society into direct contact with new phenomena. Transport investment brings greater opportunity to extensive areas most in need of it. If freight rates and passenger fares are reduced substantially, transport can, and in most cases does, stimulate use—as the case studies amply demonstrate. At the same time an acute awareness of new capabilities is communicated to a large number of people, possibly more than from any other form of investment.

The point of this kind of classification is to emphasise a different set of options from the ones traditional theory distinguishes and to stress the point that investments differ in the extent to which they affect both attitudes and technical productivity. Yields from an investment that affect mainly productivity are specific and quantifiable and may be substantial and immediate. On the other hand, the payoff from education, for example, is diffuse. It may also be substantial, but it is normally remote in time and nonquantifiable. Since it is agreed that for sustainable economic growth attitudes must change and efficiency of employed resources must rise (the two being related in an as yet ill-understood fashion), economic planners in reality have a much broader range of investment and expenditure options than that implied in the traditional rate-of-return calculus.

Transport and accessibility. Since our main concern in this study is with transportation investment, it is important to stress the degree to which the several forms of transport influence both attitudes and directly productive activities. In general, the extent of awareness of new investments and their potential noneconomic effects are a direct function of the degree of accessibility to them. Within the range of possible transport investments we have at one extreme pipelines which have a geographic dimension but have few effects of the indirect type indicated above. Indeed, a pipeline is better viewed as part of the investment in the industry it serves; it is part of a directly productive activity. At the other extreme is investment in road right-of-way.

Because of a higher degree of accessibility, a road permits use of relatively small and inexpensive units of capital (trucks, buses, cars) under independent ownership without serious economic penalty, since the evidence tends to support the belief that there are no economies of scale in motor transport. A qualification is needed here, however. In many underdeveloped countries there is a general absence of ancillary facilities which requires producers to supply these themselves. Thus, in motor transport there may be scale economies where substantial investments are needed in maintenance and repair facilities but which are not worthwhile for companies or individuals having only a few vehicles. There is little need for, and less possibility of, large-scale, remote, or alien ownership of trucking facilities than is the case with other modes of transport. This permits a greater number of small, local entrepreneurs to enter the industry and provides experience in management that may be widely shared and readily grasped. Motor transport has few of the problems associated with persistent income remission to other areas or resentment of a foreign-owned enterprise. Repair and maintenance of vehicles is not technically difficult and can be learned quickly by almost anyone willing to make the attempt. The ability to transport small shipments more efficiently than other modes of transport is especially important in the early stages of development, when trading is highly individualised and individual sales are relatively small. The greater number of participants in the trucking industry implies a more competitive outcome with the result that cost reductions are passed on to shippers to a greater extent than for other, more monopolistic transport enterprises. There is also contact with other types of business which may broaden the trucker's horizons and induce entry into new fields.

There are other advantages of a political and social nature. Centralised control of the highway system is not required to the same extent as in the case of rail, water, air, and pipelines. This permits a higher degree of local participation in both construction and maintenance. Local participation has its special problems, but it imparts experience in administration and control. Socially and culturally, road transport permits, and usually generates, a higher degree of personal contact than trains that pass or planes that fly overhead. As Haefele puts it, 'the combination of rail and trail holds far less promise for social development than does a road net. . . . Trains go by and trucks stop—an essential difference when viewed as carriers of culture instead of freight' (Haefele, 1963b). Thus the technology of road transport is such that it has greater potential for involving more people in a wider variety of endeavours than any other form of transport. It therefore has an influence on attitudes and abilities that cannot be captured in any calculation of net benefits. In the final analysis this may be more important than the direct economic benefits.

Others have pointed out the possible 'teaching' effect of motor carriers. A statement from the Pan American Union argues that road transportation is

. . . a medium that can be organised into small companies, thereby helping to create . . . a group of entrepreneurs worthy of consideration. In this connection, the carrier with small resources is distinguished from the small tradesman in that, whereas the latter is concerned only with the use of working capital, the former, because he is using fixed capital, has to cope with the more complex problems relating to depreciation, obsolescence, and maintenance. That is why the small carrier has been assigned considerable importance as a future industrial entrepreneur (Pan American Union, 1963).

The implication is that the greater the accessibility or openness and the more people directly influenced by the facility, the greater the probability of development, so long as costs of transport are substantially reduced. This does not mean that rail or other facilities designed to exploit a large mineral deposit or extensive plantations do not have significant effects. Rather, in such cases, much of the railroad investment is better construed as part of the overall investment associated with exploitation of a relatively localised productive activity.

Classification of transport investments

Lumping all transport capacity together under the heading of social overhead capital may be seriously misleading. Some portion of each form of transportation is specific in the sense that in reality it is geared to a particular industry; its developmental potential is then intimately associated with the industry in question. In such cases, it is no more meaningful to talk about the relationship of transportation to economic development than it is to refer to the relationship of any industry to growth. It is frequently suggested that transport facilities serve a wide variety of industries and it is this aspect that leads people to regard them as a social overhead. Indeed, this leads to the common-carrier obligation. On an aggregative basis this is no doubt valid, but in the process of disaggregation, the validity of this aspect of transport is considerably reduced. In an underdeveloped economy with a small undiversified manufacturing sector and a large agricultural sector specialising in one or a few crops, the transportation facilities are bound to be far more specific and less social. At the same time, the carriers function less as common carriers regardless of their legal status. In general, the lower the level of economic development, the higher the degree of transport specificity. This is true for modes that in other more developed and diversified economies are in fact, as well as by law, common carriers.

In short, it seems preferable to treat much of transport capacity as part of the main industry it serves, especially in underdeveloped economies. An oil pipeline is an obvious example, but a railway or road that passes through barren territory to connect an isolated natural resource to a local market or port is not very different. The social nature of any transport facility increases as the degree of resource isolation diminishes and the territory along the right-of-way improves; that is to say, as the level and extent of potential or actual economic development rises. Generally the longer the highway, the less specific it is, depending on the nature of the territory along the right-of-way, which will influence the extent and diversity of possible ribbon development. Likewise, the higher the level of economic development, the less specific the highway.

As soon as a distinction is made among types of transport in terms of degree of attachment to a particular industry or product, it can be decided whether it is preferable to lump the transport investment with the industry, then examine the developmental impact of that industry, or asign the investment to transport in general. As Cootner puts it, '. . . instead of lumping all railroad investment in social overhead capital, we can treat the construction of transcontinental railroads separately from investment which involves short spur lines to serve additional plants at lower cost, or double tracking, or new equipment. A new farm in a settled area need not be treated as identical with a farm on the frontier which depends on the construction of a railroad to be profitable' (Cootner, 1963).

Roads classified as merely permissive and responsive to already established trends represent part of the investment in the particular industries whose activity induced their construction. In other, non-causal situations, road construction, being responsive to already rising profit prospects, is uniquely associated with them. As far as awareness in these cases is concerned, accessibility and numbers of people, while still important, are not the strategic factors in inducing favourable economic responses. The awareness was already there. In the other cases, these factors are much more significant as inducements to change. Of course, the degree of attachment to a particular industry or type of economic activity would naturally be substantial at the outset. If further development follows, this specificity would be expected to decline in most cases. In the case of road transport, where the motive equipment itself is not generally specialised nor irrevocably committed to specific commodities or regions, there is much less long-range connection, Trucking firms established in response to, say, a sharp rise in cotton production might merely turn elsewhere in the event of a subsequent contraction. The ability of other modes to do this is more restricted.

The problem of entrepreneurship. Aside from the extent of aware-ness, which may be taken as a direct function of numbers of people affected and degree of access, there is the further problem of the kind of people influenced. This raises the issue of entrepreneurship and its distribution among both geographic areas and population groups according to ethnic distinctions, income, or education. Few cases present other than very impressionistic evidence on any of these points, and indeed there is no general agreement on the distribution of entrepreneurial talent.

In the literature on development there are frequent references to 'pariah entrepreneurship', which attribute great significance to alien minorities, such as the Chinese and Indian traders in Southeast Asia, the Lebanese and Indians in Africa, and the Jews in Western Europe and the United States during various periods of history. The uni-versality of such pariah entrepreneurship and its peculiar incidence among minority groups is often debated.

In many cases, transport facilities directly affect rural environ-ments, and there is often no apparent ethnic or cultural divergence among those living in or near the affected localities. There are doubt-less important differences in income distribution which may correlate in some fashion with both willingness and ability to respond to eco-nomic opportunity,† and good evidence of this has appeared. It seems reasonable to assume that the receptivity of and responses to new economic opportunities does not differ significantly in many instan-ces. If this is a valid inference, it follows that as far as the response is concerned, we may interpret this strictly in terms of 'awareness', which we have already equated with accessibility and numbers of people.

Yet there remain nagging doubts about such a cavalier dismissal. The fact that few specific examples occur does not prove uniformity of attitudes. Furthermore, the different effects earlier ascribed to the technologies of rail and road may not be completely technological

---

† See G. M. FOSTER, *Traditional Cultures and the Impact of Technological Change*, New York 1962, who argues that the people who are most receptive to new economic opportunity 'are neither at the top nor the bottom of the local socio-economic scale', p. 172; E. E. HAGEN, *On the Theory of Social Change*, Homewood, Ill. 1962, p. 30; and MILLIKAN. F. M and D. L. M. BLACKMER, *The Emerging Nations, Their Growth and United States Policy*, Boston, Mass. 1961, p. 38, also suggest something akin to Foster's view.

after all. In almost all underdeveloped countries the railroads are owned by the government with key positions held by members of the dominant ethnic group. In some countries, access to positions of responsibility is denied to minority groups. In the realm of transport this encourages the latter, who frequently are far more aggressive, to enter the trucking business where individual enterprise is more feasible anyway. There is accordingly in some countries an ethnic distinction in terms of ownership and operation of different modes of transport that correlates with variations in the degree of initiative, aggressiveness, and success. Some part of the observed developmental impact of road versus rail is doubtless attributable to this kind of ethnic phenomenon, especially in Thailand, and possibly, although to a lesser extent, in parts of Africa and Latin America.

The availability of finance

It is one thing to argue that knowledge of economic opportunity is intimately related to accessibility and numbers of people. It can be done by assuming that entrepreneurial talents are randomly distributed, or that in any given population the proportion of those responsive is roughly comparable. It is quite another matter to suggest that the ability exists to make the necessary investments either in providing transport or in expanding nontransport capacity. In the transport sector, the extension of service is frequently undertaken by nonresidents of the area affected. Similarly, since the resources in the region may be owned by nonresidents, the expansion of regional productive capacity likewise depends on the awareness and financial position of extraregional entrepreneurs. In Nicaragua, El Salvador, and Bolivia, there is direct evidence that much development was due to nonresident, either public or private, responses to economic opportunity. A simple relationship between numbers of people in or near a region and response is not to be expected. Yet in every economy, regardless of how poor or underdeveloped, there are always some people with the means to respond to new opportunities wherever they may be located. As in the case of public investment, the crucial consideration in such cases is the productivity, private or social, of the opportunity created by additional transport relative to other alternatives. Without a knowledge of the entire set of investment outlets, it is impossible to predict the importance of nonresident or exogenous responses in a particular region.

Furthermore, it is unlikely that the total response will come from such sources, although in some instances it may be very significant in inducing further development of the area along lines suggested by Schumpeter's 'herd of imitators'. While the number of people aware of new opportunities and capable of responding may be widely dispersed geographically, it is frequently not possible to account for this in any general sense. The model thus relies mainly on the variables of accessibility and population affected, as measured by some index of regional population. That variations may exist due to non-resident participation is freely acknowledged.

But in a more general developmental sense, the availability of finance, like so much else, is permissive only. If those possessing the liquid capital choose to purchase land for speculation or to acquire existing assets or foreign securities, there will be no development stimulated in the country where they live. What is required is a set of arrangements designed to induce productive use of supernumerary income or wealth. If governments invest in transportation facilities in the hope of developing particular regions, it is then incumbent upon them to create favourable conditions for success. Governments may do this themselves or, if this is not possible or desirable, provide inducements that make private entrepreneurs respond in the desired fashion. The latter method may take the form either of making unattractive the use of funds that have no productive value to the nation or region or of increasing the profitability of private investments of the productive type. Certainly the use of new highways should not be restrained by high user charges, heavy taxation, or restriction on imports of vehicles, parts, or other capital equipment that is needed for effective exploitation of whatever new potential is created. Bold, daring entrepreneurs with wealth may devote their energies, time, talents, and money to a wide variety of activities which do almost nothing to raise the net national product but which are privately rewarding in prestige or money. As has been said, 'What may be a source of income creation for the individual need not be a means of income creation as seen by the community at large' (Liebenstein, 1957*b*). The function of any public policy not totally committed to a goal of *public* enterprise is to create those arrangements that made 'private vice' coalesce with 'public virtue', as Adam Smith long ago argued. The mere existence of economic surplus does not dictate that it will be used productively in an economic sense. Much depends on the set of options, and their costs, available to those with the ability to undertake them.

The magnitude of benefits

The problem of determining the amount of potential benefit has been discussed earlier. At this point it is worth stressing that unless potential benefit is large compared to existing alternatives, it is unlikely to evoke much response. Regardless of motivation, little can be expected in the way of development unless relatively large profit prospects are made available. In this sense, response to economic opportunity is closely related to its amount.

Disturbance of existing institutions

It has frequently been suggested that investments that involve the least change in institutions or, what is the same thing, investments that are readily adaptable to present techniques, abilities, and incentives, have the greatest likelihood of success (Foster, 1962). A. E. Kahn argues that 'modest projects which employ relatively little capital and attempt . . . a minimal disruption of settled habits of thinking and living are more likely to succeed than those which involve a mass frontal assault on non-western patterns of culture' (Kahn, 1951). On the other hand, without a change in institutions or even in people themselves (Leibenstein, 1957*b*; Millikan and Blackmer, 1961; Hagen, 1962), there are few prospects of achieving sustainable growth. What is required, therefore, is something intermediate between a massive assault on culture and those investments that leave all else completely unchanged. This has direct relevance to road transport. In almost every case, the accessibility of roads, the ease of entry into trucking, as well as the prestige aspects of vehicle ownership, largely account for the sudden growth of motorised for-hire transport which tends to insure that lower costs of transport are passed on to users. In the final analysis it seems to be the rate of required adaptation that is crucial, and this is related to the extent to which efficient exploitation of new economic opportunity required a radical departure from existing practices. Sudden changes are apt to be resisted strenuously. Since truck transport is less different from most forms of traditional transport than railroads or aircraft and the operating skills required more readily learned, the impact is less likely to evoke resistance. This does not rule out the possibility that a dramatic challenge to existing institutions may be beneficial in the sense that it evokes positive responses by creating tensions that necessitate some activity. It is not inevitably the case that such responses are merely those which strengthen prior attitudes. In this instance, as in so many others discussed in the present chapter, the ability to predict behavioural responses to particular events is seriously deficient. Yet to ignore the effects of attitudes, as previously suggested, is to leave the matter grossly one-sided. The economist is thus torn between saying nothing in this area or attempting some crude generalisations. I have chosen the latter in the hope that behavioural naïveté may be heuristic.

There is another aspect to the issue of institutional disturbance—the reaction of those whose economic and even social position may be undercut by the new transport capacity. If such groups are unable to impede the use of the new facility, the probability of its success is sharply increased. Yet resistance to technological change has been manifest in all societies in the past and has achieved varying degrees of success over fairly long periods of time. There is no reason to expect that such resistance will be any less vigorous in the future, especially in underdeveloped countries, where neither the range of alternatives nor the resources for compensation are very substantial.

Conclusion

The general explanation of the divergent results obtained from case studies is to be found in the extent of economic opportunity created. Following are some implications with respect to the economic analysis of development.

(1) Investment options might usefully be analysed in terms not only of their direct economic payoff but also in terms of their influence on attitudes. This is relevant to the manner in which rates of return, even using the social marginal productivity concept, are typically computed. Furthermore, in every case, the actual results are correctly ascribed to a cluster of investments, ignoring policies, natural resources, and attitudes. This means that attributing any portion of the increased production solely to the highway is not only spurious, but regularly overstates net benefits relative to costs. *All* of the investments must enter the denominator, and any attribution of the total productive result to one of them is as economically unsound as finding the cost of each of two or more joint products.

(2) The educational and other spillover effects of road transportation appear to be greater than those of other modes of transport. This is especially significant at low levels of development.

(3) All transportation capacity is not 'social overhead' in any meaningful sense of that elusive phrase. Rather the social nature of transportation depends on the extent to which it is specific to a particular industry or generally available to a wide variety of industries or has educational or spillover effects.

(4) The issue of ethnic distinctions should not be avoided by economists out of a desire not to offend or for any other similar motive however well intentioned. The Chinese in Malaysia, for example, are important in imparting the peculiar dynamism to that country. To ignore this as a fact of economic as well as political and social consequence, is to create a distorted image of reality already blurred by the usual economic approach to growth.

(5) The number of elements in the growth process that are designated as 'necessary but not sufficient' is substantial. They include the following: capital in general and transport investment in particular; appropriate psychological attitudes toward economic activity and change; entrepreneurial abilities; technical abilities and education; the legal, social and political environment; the kind and amount of natural resources. In short, the phrase 'necessary but not sufficient' has become a kind of grand developmental cliché applied to so many separate notions that it might well be expunged from the literature.

SOME LESSONS FOR POLICY

This section concerns the lessons that can be learned from case investigations so that past failures can be avoided in the future and successes improved upon. The lessons of greatest relevance are, therefore, those pertaining to preinvestment surveys.

The significance of prior dynamism

The probability of success of a transportation investment is obviously dependent on the existence of prior dynamism in the region or nation as a whole. If a particular region is growing rapidly in terms of population, output, and so forth, the probability is very great that existing transport facilities will soon constitute a true bottleneck even if there is some excess capacity at the moment. The discovery of such dynamic areas not only suggests where additional capacity should be located but also is a good indicator that heavy utilisation may be expected.

If a nation as a whole is growing rapidly, the probability of making a successful transport investment is high even in a region which is not growing so long as it possesses some reasonably good economic potential. The existence of overall dynamism implies among other things an environment in which economic opportunity tends to be sought and quickly exploited when found. Thus, in any circumstance of local or general dynamism both the need for new transport capacity and the probability of success is very great. But where there is no initial growth or development, a single transportation project cannot be expected to accomplish much. It is in this type of situation that a co-ordinated set of investments, inducements, and policies is most essential and where the prospects of success from a single project of any kind are very low. The *initiation* of growth is a fundamentally different and more difficult task than its facilitation and normally requires a more careful appraisal of noneconomic factors as well.

These considerations have obvious implications for the nature of preinvestment transportation surveys. In a dynamic context, there is less need for a comprehensive, all-inclusive economic report than would be the case in a static situation. Transport economists alone would probably suffice for the economic appraisal of transport in the former case, while more broadly trained economists and others would be needed in the latter. The degree of prior dynamism obviously conditions the length and nature of the economic feasibility report. But in every instance there is need for rather thorough information regarding soils, forests, minerals, and so on. If the region involved has already been carefully analysed in terms of these natural features, the economist's job is rendered not only easier but potentially more fruitful.

What this suggests is that the role of transport specialists, either as economists or transport engineers, is or should be secondary. In particular, detailed engineering estimates that purport to give the costs of the transportation facility in very specific terms are not essential until *after* the decision to go ahead has been made. Rough-and-ready cost estimates are good enough in advance to compare with the anticipated economic benefits. The latter must be estimated first on the basis of prior natural resource potential estimates. In terms of so-called 'economic' feasibility studies, most of which are now concerned mainly with the engineering aspects of the transport facility, this could mean a greatly simplified and less costly survey prior to the decision to go ahead with the project.

Government policy toward new highway capacity

One of the important ingredients in inducing increased production is often a sharp reduction in rates usually associated with an expansion of vehicle capacity. In other words, the coming into existence of a highly competitive motor transport industry is the mechanism whereby the cost savings in transport are passed on to producers. There are several ways in which this stimulus can be blunted or eliminated: (a) by the imposition of high user taxes or tolls, (b) entry restrictions into motor transport, (c) rate regulation or rate agreements among the firms, (d) the prohibition of, or high duties on, imports of new vehicles, (e) weight and size limitations beyond those necessary to protect the highway with adequate maintenance.

Since such taxes or tolls reduce the extent of sustainable rate reductions, it is important that they be limited to those amounts strictly associated with the revenues needed to maintain the highway and to pay interest and principal on whatever loans were incurred in its construction. Alternatively, since the developmental impact of rate reductions is generally high, a tax on land values along the right-of-way, which generally rise with improved transportation, might be preferable on economic grounds and could be set at such a level that at least interest and principal on the highway loans involved could be covered. User taxes could then be limited to amounts necessary to finance those costs directly associated with use, namely maintenance and repair not caused by weather or other natural phenomena.

As a general rule in most underdeveloped countries, there should be no restriction on entry at all unless a definite safety need emerges. We have noted the substantial learning effect of widespread participation in motor transport, which suggests that the natural ease of entry should not be reduced without compelling reason. Furthermore, the more firms in existence in the absence of rate agreements, the more likely that competition will exert a downward pressure on rates. Similarly weight and size limitations for vehicles should be limited strictly to the capacity of an adequately maintained highway. The latter is important since failure to maintain highways adequately increases the apparent damage done by heavier vehicles. Such damage, which encourages imposition of stringent restrictions, is more properly related to inadequate maintenance.

Rate regulation of motor transport requires a degree of administrative overhead that is out of the question in most underdeveloped countries. Public rate regulation generally serves chiefly to prevent rate reductions and thus in developing countries may blunt the incentive to expand production in the area. Where private rate agreements emerge among the various firms supplying transport, these should be eliminated. Motor transport markets seem to be effectively competitive in many underdeveloped economies, even where knowledge of operating costs is lacking and where practices with respect to depreciation, maintenance, and repair leave much to be desired. For example, Farmer concludes, in two cases investigated, Lebanon and eastern Saudi Arabia, that '. . . without creating vast government bureaucracies controlling transport, the Lebanese (and Saudis) have evolved a workable transport system free from large or clearly unjust inequities' (Farmer, 1959, 1962).

Because truck markets seem workably competitive, and there is a scarcity of literate people who could effectively regulate such an industry, the opportunity cost of establishing, administering, and enforcing a detailed set of regulatory constraints is particularly high. As far as developmental loans for transport are concerned, a general policy favouring entry and rate freedom in motor transport might usefully become part of the contract conditions when highway construction is under consideration.

The problem of import restrictions or excessive duties on vehicles is closely associated with existing vehicle supply and the foreign exchange position. If a highway is to be built and if the present vehicle supply is inadequate to meet the expected increase in output, then the whole question revolves around the availability of foreign exchange. If there are difficulties here, vehcile imports may have to be curbed. But there is still room for manoeuvring in the sense that private passenger cars can be prohibited while trucks may be imported with few restrictions up to the numbers considered necessary.

Regular analysis of past investments

There should be initiated a series of case studies in various parts of the underdeveloped world. In the first place, it is important to check the actual results of any major investment against those anticipated beforehand. If there are serious discrepancies, an analysis of why they occurred is important to improve future decision-making. Follow-up studies of particular investments provide the only reliable, factual guide to what may reasonably be expected under given conditions in the future. This is particularly important for lending-agencies continuously required to make project appraisals. Yet there is no systematic provision for detailed reporting on what actually happened after a major transport improvement nor a careful evaluation of why certain things happened and other things did not. If any significant improvement in prior appraisal of transport projects is to take place, a continuing collection of the results of past projects would seem to be indispensable.

As already argued, a case study approach is a healthy and, in my opinion, long overdue supplement to the heavy reliance upon national aggregates. More attention to regional and local situations by economists is clearly warranted in countries yet to develop a high degree of national integration and typified by islands of relative isolation and independence. Furthermore, we have much to learn about the growth process and studies which seek to explain why the effects were as they were, add considerable insight into the appropriate set of conditions whose concurrence would raise the probability of success. A careful study will necessarily develop data or make estimates that would not otherwise be available and would thus help to reduce the size of the 'information gap'. Indeed, a continuing analysis would provide the basis for a number of time series of economic data at the regional level.

EPILOGUE

Analysis of transportation is an exercise in applied economic analysis. Yet separation of transport into a distinct area of study has generated the appearance of uniqueness which has in turn supported a belief in in what Fogel calls the 'axiom of indispensability' (Fogel, 1964). The present study suggests that there are, in fact, few magical properties in transport investments that warrant the excessive attention frequently paid to them. Transportation is merely another industry or industries; transport investment is like any other and should be judged on grounds applicable to other forms of economic activity and capital formation. The role of transport investment in economic growth is similarly not unique. Transport investment is no more an initiator of growth than any other form of investment or deliberate policy. Under some conditions, it may turn out to be strategic but the same can be said about any specific investment or policy. The essential message is that policy-makers and analysts take a more agnostic view of transportation operations and investment. The plea for more case studies of transportation and development is designed to isolate the conditions under which transportation may in fact be strategic. If recent studies have succeeded in tempering some of the ill-founded enthusiasm with which investments in transportation facilities are made, they will have served their purpose.

REFERENCES

COOTNER, P. H. (1963). Social overhead capital and economic growth, (ed.) W. W. Rostow, *The Economics of Take-off into Sustained Growth*, London and New York, p. 267.

FARMER, R. N. (1959). Motor-vehicle transport pricing in Lebanon, *The Journal of Industrial Economics* (July), p. 205.

FARMER, R. N. (1962). Inland freight transportation pricing in eastern Saudi Arabia, *The Journal of Industrial Economics* (July), pp. 174–87.

FOGEL, R. W. (1964). *Railroads and American Economic Growth; Essays in Econometric History*, Baltimore.

FOSTER, G. M. (1962). *Traditional Cultures and the Impact of Technological Change*, New York, p. 145.

HAEFELE, E. T. (1963a). Road construction as a means of developing areas served, Document No. 57, *Ninth Pan American Highway Congress*, Organization of American States, May 6–18.

HAEFELE, E. T. (1963b). Ibid., p. 7.

HAGEN, E. E. (1962). *On the Theory of Social Change*, Homewood, Ill., p. 30.

HIRSCHMAN, A. O. (1962). *The Strategy of Economic Development*, New Haven, Conn., p. 77.

HOSELITZ, B. F. *et al.* (1962). *Theories of Economic Growth*, Glencoe, Ill., p. 242.

KAHN, A. E. (1951). Investment criteria in development program, *Quarterly Journal of Economics* (February), p. 51.

LEIBENSTEIN, H. (1957a). *Economic Backwardness and Economic Growth*, New York, chapter 15.

LEIBENSTEIN, H. (1957b). Ibid., p. 113.

MILLARD, R. S. (1959). Road development in the overseas territories, *Journal of the Royal Society of Arts* (March), p. 275.

MILLIKAN, M. F. and BLACKMER, D. L. M. (1961). *The Emerging Nations, Their Growth and United States Policy*, Boston, Mass., p. 38.

PAN AMERICAN UNION (1963). *General Problems of Transportation in Latin America*, Washington, p. 31.

ROSENBERG, N. (1964). Neglected dimensions in the analysis of economic change, *Bulletin of the Oxford University Institute of Economics and Statistics*, vol. 26, no. 1 (February), p. 61.

NOTES ON CONTRIBUTORS

GAUTHIER, HOWARD L.
  Professor of Geography, Ohio State University

TAAFFE, EDWARD J.
  Professor of Geography, Ohio State University

MORRILL, RICHARD L.
  Professor of Geography, University of Washington, Seattle

GOULD, PETER R.
  Professor of Geography, Pennsylvania State University

HOYLE, BRIAN S.
  Lecturer in Geography, University of Southampton

RIMMER, PETER J.
  Lecturer in Geography, Australian National University, Canberra

STANLEY, WILLIAM R.
  Professor of Geography, University of South Carolina

MILLER, FRED
  Professor of Economics, Oregon State University

HAY, ALAN
  Lecturer in Geography, University of Sheffield

O'CONNOR, ANTHONY M.
  Lecturer in Geography, University College, London

HILLING, DAVID
  Lecturer in Geography, Bedford College, University of London

REICHMAN, SHALOM
  Lecturer in Geography, The Hebrew University of Jerusalem

WILSON, GEORGE W.
  The Brookings Institution, Washington, D.C.